CITY COUNTRY MINERS

Ode to a Telephone Pole

Fatal straight heliotropism.
Too bad, but
it's the crooked tree that
keeps its limbs, alive still
in the woods you came from.

—*John Enright*

Arthur Okamura

CITY COUNTRY MINERS

some northern california veins

edited by Michael Helm

City Miner Books
Berkeley, California

Other City Miner Books:

Bump City: Winners and Losers in Oakland by John Krich
ISBN 0-933944-01-2 $3.95

Letters of Transit, poems by Frank Polite
ISBN 0-933944-02-0 $3.95

Snap Thoughts, poems by Michael Helm
ISBN 0-933944-00-4 $2.95

Credits:

"What the Bay Was Like," © 1976 Malcolm Margolin, is reprinted from *City Miner Magazine* #2.

"Passion's Duration," © 1980 Thomas Farber, is reprinted by permission of the author from *Hazards to the Human Heart* (Dutton, 1980).

"Bed-Time Talk Between Two Six-Year-Olds" by Frank Polite is reprinted from *Letters of Transit* (City Miner Books, 1979).

"Home on the Sonoma Range" by Leonard Charles, "The Mixed Curse of the Dusky-Footed Woodrat" by Jim Dodge, and "This Year the River Had Two Mouths" by Linn House originally appeared in *Upriver/Downriver* (Box 390, Cazadero, CA 95421. Subscriptions $6/yr).

"Reconstituting California" by Jack Forbes is excerpted from *Raise The Stakes: The Planet Drum Review* #2 (Memberships $10/yr, Box 31251, San Francisco, CA 94131). "Raise The Stakes" by Peter Berg is from issue #1.

"Interview with a Gentleman Farmer" by Bruce Boston is reprinted from *City Miner Magazine* #1, 1976. "Water Piracy" by Michael Helm is reprinted from *City Miner Magazine* #15, 1980. "On Being Our Own Anthropologists" by Peter Berg is reprinted from *City Miner Magazine* #14, 1979.

"American Dream" by Lenore Kandel was previously published in *For Rexroth—The Ark #14* (Box 322, Times Square Station, New York, NY 10036.

Copyright ©1982 by City Miner Books

All rights reserved. No part may be reproduced—except for brief quotes for review purposes —without the prior written consent of the publisher. Individual contributors may reprint their work in any volume of their own collected works, provided that credit is given to *City/Country Miners*. Proceeds from any other reprints will be shared 50-50 between the publisher and the individual author. Published by City Miner Books, P.O. Box 176, Berkeley, California 94701. Printed in U.S.A.

ISBN 0-933944-03-9

Typesetting by Ampersand Typography.
Layout & Design by Nancy von Stoutenburg.

TABLE OF CONTENTS

RELATIONSHIPS
The Ticket . . . *by John Krich* 9
The Mixed Curse of the Dusky-Footed Woodrat . . . *by Jim Dodge* 15
Zen and the Art of Shaking . . . *by John Lowry* 27
Unwed Mother . . . *with Nancy von Stoutenburg* 25
Unconsulted Father . . . *with Stephen Anon* 29
Working with Single Mothers . . . *with Barbara O'Hara* 33
More Loose Leaves from the Little Black Book . . . *by Jennifer Stone* 45
Fables . . . *by Jeffrey Zable* 52
Situation NO. 2 . . . *by Daniel Roebuck* 55
Passion's Duration . . . *by Thomas Farber* 57
Riding Leathers . . . *by Deborah Frankel* 63

POETRY
Barbara Suszeanne 14, 80
Chris King 24
George Griffin................. 43
Tom Hile 44
Lenore Kandel 58, 85, 90
Frank Polite.................... 66
F.A. Nettlebeck................ 70
Andy Brumer 72
Alta 74
John Mueller 79, 81
William Garrett 82
Bruce Hawkins 86
Jack Hirschman................ 88

GRAPHIC ART
Arthur Okamura 2, 29
Nancy von Stoutenburg 9, 25
Ed Buryn ..... Photos Throughout
George Griffin................. 42
Alison Palmer 45
Brent Richardson 55
Trina Robbins 71
Joel Beck 73
Kristin Wetterhahn........... 78
B.B. Simmons 80

PLACES
Northern California . . . *with Raymond Dasmann* 93
Letter from Mendocino . . . *by Susan Pepperwood* 100
What The Bay Was Like . . . *by Malcolm Margolin* 107
My San Francisco . . . *by Stephanie Mills* 114
The Sausalito Houseboat Community . . . *with Piro Caro* 125
Mordecai of Monterey . . . *by Keith Abbott* 133
Crossing the Trans-Tehachapi Highway . . . *by Peter Coyote* 141

POETRY

Don Keefer	99
Jeffrey Wilson	99
Chris King	106
Haniel Long	111
Tom Hile	112
Stephen Kessler	120
F.A. Nettlebeck	124
Fruund Smith	131, 139
Janet Chandler	132
Jack Forbes	145

GRAPHIC ART

Joseph Carey	100
Arthur Okamura	107, 119
Kristin Wetterhahn	112
Phil Frank	122
Kate Drew Wilkinson	125
Brent Richardson	141

WORK

Harvesting the Trash . . . *with Bob Beatty*	155
Home on the Sonoma Range . . . *by Leonard Charles*	163
The River Had Two Mouths This Year . . . *by Linn House*	174
Interview with a Gentleman Farmer . . . *by Bruce Boston*	187
Private Branch Exchange . . . *by Lucia Berlin*	190
Woodshedding in the Performing Arts . . . *by Joan Schirle*	203
Midwifery: Who's Got the Power . . . *by Elizabeth Davis*	208
How To Play Hard Ball . . . *by John Lowry*	212

POETRY

Gary Gach	161

GRAPHIC ART

Joel Beck	185
Mort McDonald	186
Arthur Okamura	208
Ed Buryn	Photos

POLITICS

Frank's War . . . *with Frank B. Kiernan*	215
Neighborhood Crime Prevention . . . *with Ernest Callenbach*	221
Trust The Land . . . *by David Prowler*	228
Water Politics . . . *by Michael Helm*	234
On Being Our Own Anthropologists . . . *with Peter Berg*	245
Reconstituting California . . . *by Jack Forbes*	250

POETRY

Ty Hadman	220
Lenore Kandel	225
Jack Hirschman	233
Peter Berg	249
Cliff Eisner	243
Frank Polite	253

GRAPHIC ART

Joseph Carey	215
Brent Richardson	221
Editor's Note	255
Contributors	256

RELATIONSHIPS

*All day I worked
with the lingering
scent from your thighs
upon my must
stash.*

—Michael Helm

Nancy von Stoutenburg

THE TICKET

John Krich

Never trust a cop who wears perfume.
   I was out on my customary cruise, doing basic research for the groin, when a squad car turned its red light on me. Lowering my window, I discovered I was dealing with a uniformed pixie. And it wasn't such an easy discov-

ery: hardly any tits under the badge sewn to her khaki pocket, hair shaped so two points could poke from her cap like homegrown sideburns. Then I picked up the officer's scent. Maybe it wasn't perfume. Maybe it was aftershave.

She practically stroked my lap reaching for the license. What was she getting so excited about? This was just routine harrassment: the beginners' lesson taught in cop school. But this might have been her first chance to try it out on the street. She sure knew how to get back to her console and run my name through the big computer in the sky. Miss Piggie, or Missus Piggie, or maybe Ms. Piggie, came back looking glum. I knew I was clean.

"Do you want to tell me exactly what you were doing in this vicinity?"

"Exactly?"

"That's right, sir."

"Exactly nothing."

"Then why were you slowed to five miles an hour?"

"Everyone's got a right to look."

"You were just looking at three in the morning?" This cop was quite a wit. She laughed at her own jokes—another procedure they teach in cop school.

"Best time to look," I matched her. "Less smog in the air."

"Then I suppose you saw we've got a prostitution problem on our hands."

"Oh really?" I should have told her I wished it was on my hands.

"Don't sound so innocent. As long as there are men like you, there will be girls on this street."

I should have asked her what she meant by "men like me." Instead, I got fancy: "I tend to see it the other way around....I mean, which came first, the sperm or the egg?"

"Okay, Mister Philosopher. Let's see your lights."

Lesson number two from cop school. She was putting me through the whole bit. She was gonna make sure my vessel could float.

"Sir, I'm afraid one of your brake lights is out." Afraid, shit. The pixie was in pixie heaven.

"You don't say?"

"See for yourself."

"That's okay. I believe you. You see, some people don't take care of their wheels the way I take care of mine." If there was anything I'd learned from growing up in this burg, it was that you were safe if you had functional wheels. The town was nothing but one big amusement park ride.

"Then this isn't your vehicle?"

"My old lady's."

Now the cop was really swinging.

"So you're out on the boulevard, looking over the prostitutes, in your wife's car."

"My girlfriend's car. Mine's in the shop."

"And does your 'girlfriend' know where you are?"

"She knows I'm just looking."

Was this cop gonna write me a ticket or a morality play?

"Can I see the registration, please."

"Sure thing." But I was getting less sure. It had to be in the glove compartment, but the glove compartment felt like it was full of confetti. Lots of maps in there, lots of kleenex, but not even a pink slip. Served me right for having an old lady who was less together than I was.

While I kept digging, the pixie sent for reinforcements. She must have thought I was reaching for a piece. Her partner was no android—there couldn't have been two like her in the force. He was another genetic piece of cake: black and half-baked.

"We don't need your high school diploma, son." Supercop to the rescue—and a real charmboat, too.

"My old lady's got it at home, probably in the drawer to her sewing machine. But it's her car. I can call her up right now if you want..."

The blessed couple went back to their front seat and let me sweat some more. Three in the morning, with the window rolled down, and I was sweating.

"You want me to call her?" They were back.

"You've made enough trouble for the poor girl as it is." The pixie was all heart. "We'll just cite you on these violations. No papers, no lights. That's a heck of a way to drive around this part of town."

"Now I know." There was really no reason to sweat.

"And we don't want to see you on this beat anymore."

"There's no law..." Not that I cared about laws. "It's a free country..." Not that I cared about the country. "I can go where I like..."

"You can also go to jail," the black intoned. "We've got a big enough problem down here without you."

I should have said that the problem was in the eyes of the beholder. Instead, I admitted, "Yeah, it's a big problem. A lot bigger problem than my busted brake light."

"Listen, don't think we don't know what you were up to."

"Okay, I won't. I don't even *know* what I was up to."

"Sure thing, buster."

"Look! Here...here's my wallet. You want my wallet, too? It's got five bucks in it. How far can I go on a fiver?"

"Plenty far with these freaks."

"How would you know?"

The pixie was beaming to hear all this he-man talk, but the black cat turned his back on me. I guess he was tired of being my straight man. He helped the pixie tear a yellow sheaf from her warrants pad. A carbon fluttered to the curb.

"You've got fourteen days to correct these infractions. Now go on home."

"Yes, m'am. Thank you, officer." Roll up the window, turn the key, you are safe. It's back on the roller coaster, three spins for a quarter. I stuffed the ticket into my shirt pocket without reading it.

After dinner the next night, while my old lady sipped on her Darjeeling tea, I whipped out the ticket and dropped it on the table.

"You got me in a pile of trouble last night." I told her the whole truth and nothing but the truth.

"It's not my fault if you were prowling around at all hours."

"But you didn't even have the damn registration in there."

"But you got caught."

"But you had only one brake light."

"But you got caught."

This had a way of getting old fast, and besides my old lady didn't take lip the way those porkers did. That's why I was out riding, not snuggling up in bed with her.

Then she spun the ticket back at me and pointed at some scrawl near the bottom.

"What's this, kiddo?"

Like an idiot, I read it aloud: "Talk-ing to known pros-ti-tute."

"Were you?"

"They can't do that."

"Were you? I don't want you to give me herpes or something. You know, once you get it, herpes is forever."

"Relax. You can't get herpes from talking...and besides, I didn't do nothing. I was just gawking, just stimulating my appetite."

My old lady said she knew all about my appetite and I knew she was relieved for me to take it elsewhere.

"That she-cop was trying to get back at me. She figured you'd never understand."

"I don't." My old lady was grinning, so I grinned back.

"She probably figured you'd boot me out of the house."

"I should. That's what I should do....But where would I find someone else who was stupid enough to be so honest.?"

"Gee thanks. But it's not fair, is it? They got me tagged for all time." Yeah, I was like some migratory bird with magic marker on my wing. Now they could follow me everywhere. In some dumb file cabinet, in triplicate, it was on record, irreversibly, what kind of man I was. "Suspected fornicator." "Premeditated copulator." Why didn't they just put down "sucker, john, dickbrain"? Stamp me "out of control"? It was against the all-American amusement park rules. It wasn't cool. It wasn't even the truth. "You know what it is, kiddo? It's a violation of my human rights!"

"Just get the ticket taken care of." My old lady had a head on her shoulders.

"But it was your car."

"But you got caught." My old lady was a fucking bore.

The next afternoon, I got the light fixed. It was just a loose wire. I wish I could have had myself fixed. Yeah, I wish lust was just another misdemeanor. I took the ticket, along with the registration, to a highway patrolman and he counter-signed the scrip to show that I'd complied. I watched his eyes while he was doing it to see if he was reading the part about talking to a known prostitute. I didn't want him to read it. I kept waiting for him to crack some joke.

I figured I'd hear one at the county clerk's window, but the only joke they told me was "Forty-four dollars, please."

"But it's fixed."

"No matter. New fee schedule."

Forty-four bucks? For that money, I really could have had a blast on the boulevard. I could have gotten a big daisy-chain going.

For the first time in this story, I knew what came next. The only way to get out of paying the full amount was to go see the judge. Alright, I'd do it. And if that motherfucker dared to read aloud what they'd written down at the bottom, then I'd squawk back. I'd be the Patrick Henry of the prick. "Give me liberty or give me a fine!"

When I got up to traffic court, I found I had to take my place on a long line that worked its way up the rostrum down the center aisle between wooden seats. Plenty of time to rehearse my speech. Plenty of people to hear it. They'd all find out I'd been talking to a known prostitute. They'd get their chuckle and it wouldn't matter if I was right or wrong. That cop with the perfume would have her triumph.

That's when I started to chicken out. That's when I figured I'd rather just pay the money. But it was too late. I felt just like I used to in eighth-grade algebra class. I got scared the way I used to when the teacher called on me and I didn't have the answer. It was always later, when class was over, when no one was listening, that I had all the answers.

Now there was just one dude ahead of me, claiming he'd double-parked for six hours so that he could see his dying grandmother. That yarn gave me some courage. The judge was way up above me, elevated in levels of waxed veneer. He looked like Charlton Heston. He glanced at me, reading me, then down at his copy of the ticket, reading it. He saw section 114A and Section 34C. Now he had to be seeing it. He had to be spying at "Talking to known prostitute."

"All repaired now, hmmm?"

"Yes, your honor."

"Court appearance accepted in lieu of bail," he tossed toward a bailiff.

"Court appearance?"

"That's all. There's no payment."

"Thank you, your honor."

I fought my way back through the line. I didn't feel outraged or scared anymore, just tricked. I shouldn't have let them make me feel small. I should have remembered what one jerk did to you, some other jerk always undid. That was how America worked, that was how everybody stood it. "Checks and balances"—I remembered that much from Social Studies. That was how they made you feel thankful for getting out of some mess that you never should have been in for starters. I crumpled the ticket, and heading out of court, I nearly stopped short to salute the goddamn flag.

Full Moon Madness

Full moon madness
has come over me
I've been seeing
what I shouldn't see
everyone has got a moonlit tan
shaking with fever, hear the moonlit band

There's some people on the streets tonight
some come for the loving and some come for the fight
you can be crazy when the moon is high
flow with it and blame it on a moonlit sky.

If you never fell to the feeling
then maybe you don't understand
why people are dancing & singing
in time to a moonlit band

Full moon madness
has come over me
the band is playing
in soul harmony
Everyone's singing cuz they all caught the beat
hungry & sweaty in the nighttime heat.

There's some people on the streets tonight
looking for madness in the full moonlight
I know it's crazy & I don't know why
but you can flow with it when the moon is high

Full moon madness
has come over me
I get the notion that I should be free
everyone has got their own trips too,
I see their crazies and the moon does too.

—lyrics by Barbara Suszyeanne

THE MIXED CURSE OF THE DUSKY-FOOTED WOODRAT

Jim Dodge

F ive years ago I built a small studio/sleeping room in a grove of redwoods near the Sonoma coast. I disturbed a woodrat's nest while clearing the site. It was unintentional; I was rummy from a long hot day of grunt work and just didn't notice it—one of those fairly common human acts that begin in all innocence and become sickeningly conscious about half way through, when it's too late to stop. I say innocence, but it's more likely the usual combination of inattention and ignorance, and it always seems one pays dearly for such mindless acts. I certainly did. And do. For my little studio in the trees was immediately hit with the mixed curse of the Dusky-Footed Woodrats.

The first inhabitant, a psychopathic male, was living in the walls before I hammered down the last board. There have been three others since him, two females and then another male. While my purpose here is to demonstrate the great diversity of character among a single species (in this case *Neotoma fuscipes*, the Dusky-Footed Woodrat) and to suggest that the heartmeat joys of life are not taxonomic, where identification is confused with identity, but in those diverse and particular expressions of being, nonetheless some general information about the Dusky-Footed Woodrat might prove useful.

It is roughly 14-18 inches long, half the length being furnished by a tapered, hairless tail. Adults weigh from 8-14 ounces, with males usually bigger than females.

Their body coloration varies quite a bit, but usually it's grayish brown with a gray or white underbelly. The hind feet are sprinkled with dusky hairs. In my observation the males tend to be much more brownish than the females, who are a sleek blue gray with a striking white underbelly.

They live in heavy chaparral, stream-side thickets, and deciduous or mixed woods. Their range extends from northern Oregon down through California into the northernmost part of Mexico. In California they are seldom found in the Central Valley or on the eastern side of the Sierras.

Except for occasional mating forays and (in the case of females) rearing the young, the Dusky-Footed Woodrat is a solitary creature. Most established references indicate that litters are usually produced in May or June, though occasionally from October to January, with 1-3 offspring per litter.

There's a notation in *Petersen's Field Guide to Mammals* that grossly understates, "Shows ownership of house (territorial)." That's like saying sharks eat meat. Once a woodrat occupies the house you've built, you're standing on its turf. And they don't like it. Especially the males. The females seem to understand that you need shelter, though they make little effort to hide their annoyance at your presence.

Woodrats subsist by a foraging/gathering technique most accurately described as theft. Nothing is safe unless it's sealed in a glass jar or a strong metal box, and then only till they figure out how to open it. (When a Freudian-oriented friend recently visited me and noticed the collection of jars and metal boxes on my desk he said with some alarm, "Good Christ, Jim, your desk is a classic exhibit for the anal-retentive personality!" The incredulous and concerned look on his face when I blamed it on woodrats hardly encouraged me to pursue that line of defense.)

Dusky-Footed Woodrats apparently have two main "spaces" in their over-all territorial scheme: a stash and a nest. In the natural world, the stash (or "house") is made of sticks and other sturdy materials. They stash their loot in the houses. Since the woodrats who've lived in my studio already had a house, they simply stored their take behind the books on the top shelf of the built-in bookcase and, later, in a stuffed chair. I've never seen their nests, where I assume they relax between raids, and since I've looked hard, for reasons of revenge, I've concluded they're exceptionally well secured.

As an example of woodrats' indiscriminate rapacity, here's a partial list of things (most of them belonging to me) that I've found in their stashes:

Bay berries (actually nuts) gathered from nearby Bay Laurel trees
Assorted fungi, mainly mushrooms, dried beyond positive recognition
Candy bars, cookies, and other goodies
Pencils and Pens
Dice
Flashlight batteries and bulbs
C-clamp
An old toothbrush
Paintstik Livestock Marker
Book matches
A bear claw
Marbles
Thumb Cymbal
Dried skunk foot
Adhesive and cellophane tape
Address books
Coin wrappers
Bobby-pins (from Vicky, my spouse-equivalent, significant other, living together partner, main squeeze, and general old-fashioned girl-friend; she shares the room with me)
Tissue paper
Sharpening stones
Corkscrew and bottle openers
Combs
.243 casings
Stationery of all kinds
Kindling (dragged noisomely through the walls from the wood-box)
Single playing cards and whole decks
Fishing lures
Precious medicines and herbs
Letters
Feathers
Dental floss
Chainsaw files
and generally anything portable left out loose

There is some agreement among reputable sources that woodrats will "trade"—that is, if they take something of yours they will leave something of their own in return, for which reason they're also known as "trade rats." It's always been my understanding that trade involved mutual assent about the objects to be exchanged, and that generally the objects were of roughly equal value (value, of course, being an individual and species judgment). From my experience with woodrats I have determined the following exchange rates:

Woodrat Takes
Three pens
Address book
First edition of Jack Spicer's *Billy the Kid*
Entire bottle of Percadon when you've just had three wisdom teeth pulled
The last candy bar, which you'd saved for three days and were looking forward to with that keenness that accompanies discipline, all set to plunge into the yummy

Woodrat Leaves
One rat hair
A flea
Confetti
One Bay nut, wormy
One dry rat turd

Woodrats communicate through squeals and tail-whapping. Both methods produce an obnoxious noise. The squeals range from short, staccato shrills to piercing, elongated shrieks that raise the small hairs on your spine and instantly turn your brain into cold lemon Jello. The squeals seem reserved for moments of outrage, indignity, mating play, and mortal danger. In comparison, the tail-whapping is almost charming. I've never actually seen woodrats whapping their tails, but the sound itself (a flat, rapid *whap-a-dap-dap-dap-dap*) leads me to what I consider an overwhelming inference that only a long thick tail beaten madly between two boards could produce it. To my ear there are two kinds of tail-whapping. The first is questioning, about five quick whaps, tentative. The other, oftentimes forty quick, percussive whaps, expresses a crass, malicious glee in their accomplishments, like tipping a glass of water over on your manuscripts, or a particularly bald and daring theft. This latter whapping is sounded at the point of certain safety, and is somewhat analogous to the *beep beep* of the Roadrunner as he foils Coyote into cartoon mayhem.

Besides these communicative sounds, woodrats produce an amazing amount of secondary noise-heavy rustling, scratching, gnawing, nervous scurrying, and other grating distractions. These ancillary sounds are timed to coincide with the peak moments of your concentration.

The woodrats' larcenous nature is perfectly complemented by their amazing strength and dexterity. I've seen them many times snatch up a whole candy bar in their teeth and run up the face of the five-foot bookcase in less than two seconds—or roughly equivalent to you scaling a 50-foot wall with a 20 foot 4x4 clenched in your jaws—possible perhaps, but not in two seconds. However, if you *can* do it, I have the floor plans of the Lisbon Diamond Exchange that I'd be willing to show you for a small piece of the action.

The first inhabitant—or, perhaps I should be more careful and say the *original* inhabitant—of my studio was the large psychopathic male mentioned earlier. If it was his nest I'd disturbed in constructing mine, he exacted his revenge in fair and constant measure. I seldom saw him. In fact, if I hadn't pinned him a few times in the beam of my six-volt during acts of flagrant destruction, I might well have surmised my room was occupied by some bodiless demonic force.

Most woodrats gather; he scattered. What they would pilfer, he destroyed. He trashed what he touched, and he touched everything he could reach. He flung my papers and correspondence on the floor. Pushed books from the shelves. Worse, he had an utterly accurate and completely malicious sense of the personal: he ate Roethke's face off the cover of his collected works; stole my Tenderfoot pin which I'd treasured since my Boy Scout days; gnawed the leather straps off my fishing creel; thieved the antler pipe my brother had made for me; pissed on the open pages of my journal.

He usually wreaked his prime havoc between midnight and 4 A.M., though he ran an occasional commando raid during the day to keep me loose. If I was gone for a couple of days, it often took me an hour to clean up my room when I returned. Once when I returned from a week in the mountains I could hardly open the door; I seriously considered Vicky's suggestion of renting a backhoe to clean up the mess.

While the constant trashing was annoying, I tried to suffer it with goodhearted amusement, keeping in mind that since I'd displaced him to shelter myself, it was my just dues. But after eight months of tolerance, he began doing something that snapped my chain: late at night, as I lay snuggled asleep with

Vicky, he would hurtle across the room like some rodentoid banshee and tap-dance on my face. To say it woke me up is like saying it's scary to fall out of an airplane: a million volts of adrenalin hit the roof of my brain. The first few times I went berserk, raging around the room stark naked and stabbing at any noise in the walls with an icepick, totally bent on mayhem, till Vicky's laughter eased me back to my senses.

The woodrat compounded this cruelty by only scampering across my face about once a month, thus leaving the other 29 nights restless with dread. Of course, it crossed my mind early on to stay awake with my six-volt and a .22 pistol loaded with shorts and jacklight him on the spot. Unfortunately, it has been my long-standing principle that I would kill only for food or defense of my life (or the lives of family or friends), and after scouring my conscience I had to admit that a woodrat galloping across my face, while certainly odious, was not a threat to my life. Nor did the thought of eating him particularly excite my palate. Frankly, I was eager to add a bunch of exceptions, amendments, and riders to my principle, but I'm proud to say I didn't. Principles shouldn't be easy. If they were, they'd melt away into situational ethics, and we're already beggars of circumstance. I must admit, however, in the face of my righteousness, that I seriously considered *hiring* someone else to kill him. After awhile I got used to occasionally howling awake, and as I grew accustomed to it I found I could usually get back to sleep within two hours after the twitching stopped.

Almost exactly a year after he'd begun his haunting torment, Trasher disappeared. I'm sure some of the people who live here suspect I accomplished his disappearance with a slow poison, but in fact I have no idea why he vanished. Perhaps he died of natural causes (woodrats live about four years maximum), or perhaps a cat or owl nailed him. Personally, I suspect the Woodrat Spirits directed him to go punish some other nest spoiler. I was immensely relieved, of course, and surprisingly sad.

The next inhabitant followed in a couple of weeks, a young female I'll call Sugar Mama (Vicky and I always referred to all the woodrats simply as "Rat"; I've added names only to differentiate them). Sugar Mama was a high-stepping lady and a consummate thief. She liked the sugar in life. She mated with a frequency not even hinted at in reference works, and her salacious ways evidently gave her a taste for more sweet rushes because she would steal anything sweet no matter how well you thought you had it protected.

When the great Willie Sutton was asked why he robbed banks, he replied with your basic, straightforward criminal logic, "Because that's where the money is." Sugar Mama employed the same logic regarding sweets. After two months of unrelieved loss of my candy bars, cookies, and general goodies stash, I resorted to a 12"x8"x4" tin box with a tight fitting lid and a push catch that secured it. And then I put it in a cubbyhole on the shelves, smiling at the close fit. She would have to push it out, open the clasp, and pry up the lid. It took her three nights to figure it out. She looted it bare, and when you get into town only twice a month, and have something of a sweet-tooth yourself, the loss hurts. I was finally driven to a kind of perfection: as far as I know, I'm the world's foremost authority on rat-proof containers. (The best one, Vicky bought at a flea market. It's an old first-aid box that was used in Appalachian coal mines during the '30s. It is made of a thick metal I can't identify, has a plate on the bottom to bolt it down, and two strong snap-locks as closures. Until woodrats learn to use plastique, which I'm sure they will, it is the best container to protect your treasures and indulgences.)

Her libido was as limitless as her taste for sugar. The walls of my room literally resounded with passionate scampering, thrilled squeals, and the whap-a-dap of thrashing tails, which I imagined entwined in some rapturous rodent bliss not that much different from our own. Sugar Mama never played hard to get. The old schoolboy adage was true in her case: she could trip you and beat you to the floor.

Aside from her fondness for my weaknesses and her scandalous sexual behavior, Sugar Mama wasn't much of a nuisance. At first, like Trasher, she came out only when we were in bed with the lights out, but it wasn't long before I caught her peeping over the top of the books as I sat working at the desk. Every night she would show a bit more of herself, until at last she was in plain view a moment before scurrying back into the wall. And then she did something that struck me as fairly remarkable: she showed herself on the books, then turned her back and dangled her tail over the edge, twitching it to insure my attention. I rose very slowly from the chair and peeked over the books; she had her head twisted around, watching her tail with a canny eye. It was the bait to test my intentions: if they were predatory, all she would lose was a bit of her tail; if they were benign (which they were) then she could do whatever she wanted. Which, of course, she did. After awhile she ignored my presence completely. One night, hearing me unwrap a candy bar, she plopped down softly on the desk and ran over and took it right out of my hand. She was making her getaway, slowed somewhat by the weight of her load, when I reached out and snagged her tail, giving it a sharp retaliatory pinch before releasing her. She dropped the candy bar and spent the next three weeks worriedly dangling her tail over the books. It was tempting, but I let her be. She seemed to respect me after that, but she never let it become complete trust.

Sugar Mama's rampant mating finally led to a brood. I found the two newborns, pink and squeaking, in the top of a chest-of-drawers I keep outside for storing odds and ends. Although I didn't touch them or the nest, Sugar Mama moved them immediately. I was sorry I'd disturbed them. But for some reason I kept blundering across the new nests every week—in a cigar box, under the chair, in the woodbox—until finally she gave up, maybe figuring since I'd found her babies half a dozen times and hadn't eaten them, the constant hassle of moving them was greater than the risk I presented. Twice I saw her moving them. She carried one with her mouth, teeth clamped to its neck scruff, while the other clung to her underbelly.

One night not long after the young were weaned, I heard her giving them their night scurry lessons in the walls when suddenly they fell silent. Like a heavy crystal precipitating from a vacuum, the silence gathered mass. And, just barely, I heard the slithering whisper of a snake, the dry rustle of plate-scale across dry wood, then four sharp shrieks and silence.

While I'll admit to being seriously ripped at the time, and therefore acknowledge that my sensory reliability is open to fair doubt, I *know* a snake ate them.

But, I think, not all of them, because I woke one morning about a week later to find my pens and pencils gone. A new woodrat, I thought, but to make sure I left two chocolate chip cookies, freshly baked, on my desk to see if they made it till morning. They didn't make it five minutes: a young female rat, just the right size to be a surviving offspring of Sugar Mama, came charging down the bookcase and scooped up one of the cookies. However, the cookie was bigger than she was, and as she attempted to get it back up the bookcase she looked like someone trying to scale the World Trade Building while carrying a 4'x8' sheet of plywood. She kept dropping it. Finally she got nervous or frustrated and gave up. Before I went to bed I broke each cookie into four pieces and left them on my desk. There wasn't a crumb in the morning.

You don't domesticate wild animals, you tame their young. Sweetie, as I'll call her, had been aware of my presence since birth, so her natural wariness was somewhat blunted. (Also, I'd helped her with the cookies, which I like to think she appreciated as an act of kind concern. Other animals no doubt have different minds than humans, but they aren't stupid. Sweetie knew where the goodies were.)

At first she half-ass raided, as her mother had properly taught her, but it was only a matter of weeks before she came out and tried to open the candy box while I was working at the desk. Any sudden movement from me, though, and she went scooting for safety, so I began talking to her in a calm, reassuring voice, telling her how lovely she was with her fine blue-gray coat and striking white underbelly, like a deer in November, and promising not to harm her as long as she didn't unduly harass me with either theft or noise. After a few weeks of such convincings, she ventured over to sniff my hand, her nose moving like a pencil eraser receiving a slow electric shock. She must have decided I was all right, because it wasn't long before she was taking food out of my hand.

When she first started doing it I was careful not to scare her, for it was an opportunity to observe a woodrat up close. Such sensitive, elegant whiskers; such beady eyes. I also observed a number of fleas and ticks, which brought me a few hits of plague-dread and general aversion, but they quickly passed. In general, woodrats seem a much healthier group than sewer or wharf rats.

After I'd observed her for about thirty feedings, I either began some experiments (my claim) or became perverse (Vicky's opinion). Anyway, I quit letting go of the candy so easily. Sweetie was outraged at my treachery, and after a few minutes of frantic tugging and prying, bit me sharply on a knuckle. It didn't draw blood, but it left a clear white print of rodnt incisors and certainly made me let go real quick.

Those nights when there were no goodies she got pissed off. She'd hang around expectantly, give her tail a few impatient whaps, dash over and nose my hand. When it became obvious that I wasn't holding out, she would sulk off into the walls. Inadvertently, out of playful interest I'd tamed her, and I've regretted it since.

Sweetie's stay was a relative pleasure, lots of fun and minimal disruption, but I hadn't prepared her for the world she lived in. One night, after an eleven month sojourn, there was a tremendous battle in the walls. At first I thought she was mating, but when she didn't show up the next night I feared for the worst, a disposition confirmed when my possessions began to disappear. I took my flashlight to bed the next night, and when I heard the familiar scampering of woodrat feet on my desk, I snapped it on: it was an enormous male, a two-pounder for sure, shambling across the desk with my new pen in its mouth. It was clear that Bruiser, as I'll call him, had run Sweetie out of the house. Whipped her ass and threw her out. Vicky took an immediate dislike to him, and I must admit I find little to recommend him after seven months of having him in the house. He steals everything left out loose, so once a week I have to spend about an hour digging around in the stuffed chair to retrieve my belongings, and that's a drag. He is also inordinately fond of tearing up paperback books and piling them around the base of the chair to secure his stash, though the chair is only an inch or so off the ground. He is also strong enough to drag our boots around the room, which leads to a lot of stumbling and stubbed toes as we search for our footwear in the early morning darkness. He likes to chew on dirty socks, an inclination I find it hard to sympathize with. Worse, he apparently has no sense of humor. Even Trasher had a sense of humor, albeit malignant. Bruiser is just stolidly glum, marked by that petulant meanness common to bullies of all species. He doesn't even whap his tail very much. But what I like least about him is that he gets me in trouble with Vicky.

It's like this: after Bruiser had been in residence for about four months, I was sitting at the desk working one night while Vicky was in bed reading, and suddenly our long lost Sweetie came scampering down the bookcase and ran over and nosed my hand. I was delighted for a moment, called softly for Vicky to look too, but then I noticed that about two inches of her tail had been bitten off, leaving a raw stump, and her once sleek fur was now rumpled and mangy. Her fine whiskers appeared frayed, and the glint in her eyes seemed considerably dulled. I reached for the candy box to give her a treat, but she didn't wait. Instead, she scampered rather blithely across the desk and dove into the stuffed arm chair. A short, fierce battle with Bruiser ensued, punctuated by three sharp shrills of pain, and she came hot-assing out of the chair and up the bookcase without a glance in my direction and was gone in a blur. I said something like "Well, I guess that's what being territorial is finally all about," and I sort of laughed, sympathetically I thought, having experienced a few instantly decisive defeats myself. I mean sometimes you just get the shit beat out of you and that's the way it is until you think of a way to get around it. Vicky didn't agree, nor did she see any humor in it, except of the meanest sort and coldest kind, like putting piranhas in a child's bath or spitting on puppies. For a couple of weeks I thought I was going to have to move into the chair with Bruiser. Finally I agreed with her suggestion that we trap him and take him far away, but insisted that she do the trapping and moving, which to her credit she hasn't done, or not yet anyway.

Perhaps it won't be necessary. It's been very quiet in the walls the last week, and just the other night I'm sure I saw Sweetie return again and enter the chair without incident. If it is indeed the nature of things to become their opposite, perhaps the mixed curse is becoming a mixed blessing, or at least maintaining that dynamic equilibrium which, it seems is sustained by the faith of both woodrats and humans.

Zen and the Art of Shaking

    One day, out of the blue, Tom started shaking people.
'I want to change them,' he said.
    A guy with a handle bar moustache spit in his face.
    A girl with acne had an orgasm and had to be taken away.
    An old lady threw down her glasses and said, ''Hello, Lemuel.'
    A skinny guy carrying a book thought a moment and said, 'Yes, the world is spinning faster.'
    A baby smiled and blew a big spit bubble.
    In Central Park, Tom shook a tall woman with heavy brows. It was like shaking a wall. Then, placing her hands on his shoulders, she shook him until his eyes rolled in his head.
    Smiling, Tom walked away and took up sky diving.

—*John Lowry*

You sleep with
cats on your lips

They breathe
through you
at night
in bed

and the dog
in the corner
of the room

Your children
Our children

Tell me that
my body isn't
the world
outside
bound up
together
and you
this storm

I stare out
the window
in a bathrobe
and see

winter spreading
its gray lobotomy
across the sky

Hand me a glass
of water and I'll
shave my neck
and start
to conform

But you laugh instead
while the cat raises up
underneath the sheets

and I throw
the covers back

and you turn to me
smiling with the
cold thunder of
eight more lives.

—*Chris King*

RELATIONSHIPS 25

Nancy von Stoutenburg

UNWED MOTHER
Nancy von Stoutenburg, talking

I arrived at the decision to be a single mother because I wanted to have a child but didn't want to be married. I once was married. Five years in common law and two legally. After two years of really being married I felt that I wasn't doing what I wanted to do. I was doing what my husband and I wanted to do together. He wasn't doing what he wanted to do. I wasn't doing what I wanted to do. We were doing what the "couple" wanted to do.

But the question of not wanting to be married is still a separate question from deciding to have a child. I thought about that a lot. I like children. I've always wanted to have a kid and finally got to the point where I thought I could have one and did. I chose a man to father my child that I consider intelligent and genetically clean. I don't know that I would have chosen him if I had to live with him. Alan's intelligent, physically sound, and talented. I think altogether the kind of person that I would like my kid to be, except that he's crazy.

It was agreed, between Alan and me, that he could have as little or as much to do with the child as he wanted. It was his child, too. He knew that I was trying to get pregnant. There were occasions when he said, "well what if I don't want to have a baby?" And I said, "well, then don't fuck me." I was only sleeping with him at the time, although he was sleeping with other women. Before that, I was sleeping with my husband whom I had separated from. Because I still think his genes are pretty good, too. In both cases I was in love. I still love both of them very much.

I got a variety of reactions from people as a result of going ahead and having a child the way that I did. From one end of the spectrum to the other. When I first got pregnant there were lots of people, particularly men, who said, not to me, but to other people, that I was crazy to do it. And a lot of those people were people who themselves had failed in their marriages and their children have suffered for it a lot. Rick, a close friend, for example, first thought that I shouldn't do it, that it would just really be a bummer. But then, after a while, the more he thought about it, he admitted this to me: he said, "the more I thought about it, the more I realized that you are better equipped to bring up a child than most of the people that I know that are bringing up children, because of your attitude towards children." Rick told me that his initial reaction was because of ego. He didn't want to think that a woman could bring up a child without a man, that that dispensed with him and his role.

Of course, I realize that having a kid is also an ego trip for a woman. But I'm not a chauvinist. I believe that there are some men who are a lot more capable of bringing up children than some women are. I think it's strictly a matter of the individual parent on who's best to bring up a child.

I don't think that actual genetic parenthood gives anybody an unqualified right to bring up a child. When I think of Son of Sam, for example, I think, as an infant, somebody should have seen what his parents were doing to him. You know, he turned into a maniac. His parents mistreated him. And when they put him in jail they should have put them in jail too. They should have taken him away from them when he was an infant. There are people who should not be allowed to bring up children. I believe that children are born perfect and if left to grow they will grow perfect. If made, if molded, then they will become a distortion of their perfection. I think that as soon as a child is able to say that he doesn't want his parents and means it, that he should be listened to with very strict attention. Because there are very few children that will say that. And if they say it, by God, they really mean it.

I really didn't know, I didn't pretend to know, what it was going to be like to have a kid. I knew that it was going to make my life totally different than it had ever been before. And it has. I like it, I like it a lot. But, of course, I've only been doing it for seven months. Ask me again in two years, or thirteen years. I like Springsteen, my daughter, a lot. I think she's a great kid and we have a lot of fun together. She's really starting to develop a personality. It's a lot more fun than I thought.

A lot of my fears about motherhood came from what other people told me. They told me, "Just wait, you just wait, no matter what, you just wait. Wait until they start crawling, wait till they start waking up every two hours."

There are women who have, after seeing my experience, thought that it probably would have been easier for them, in a lot of respects, had they done it alone. But that's because a relationship with a man is totally different from a relationship with a child. Often times two people, who plan on having a child, don't really know each other's opinion about child rearing, about what they really expect. So that creates a potential conflict situation for the kid.

In terms of single motherhood there's a standard view, which I don't necessarily agree with, that says a little girl who has a positive experience, a good relationship with the father, then has a role model that makes it easier for her emotionally to connect to other men. And in the absence of that it might be more difficult for her. I don't know though that there will be that absence for Springsteen because there are men that she really relates to now. I actually have a tendency to get along with men, friendwise, better than I do with women. Most of my friends are men and Springsteen gets along with them very well. There are two men in particular, John and Henry, that she loves already. John is her godfather. I think that she relates to them and they relate to her in such a way that she will have good feelings about men.

I do have a lot of help in raising Springsteen from my friends. If I hadn't had such good friends I don't think I would have had a child. If you're a person who likes to be alone a lot, then you probably shouldn't have a kid.

I have also hired babysitters and been fortunate in that I've been able to make enough money to support us comfortably. It's getting tougher, though. She's getting more expensive. At first everything was given to me. I didn't have to buy anything for her. Absolutely nothing. I breast fed her so I didn't even have to buy food for her for the first three months. I don't think I spent a penny on her. Except to buy a rattle, or something, for 59 cents.

In terms of child rearing principles I pretty much worship A.S. Neill. He ran Summerhill. A lot of people say his views were idealistic, utopian. And yet, he *did* do it. And so other people can do it. I've read inspiring interviews with kids that graduated from Summerhill. After Neill died the school changed. The anti-Summerhill period started. But his idea of children is like what I said before, that they are born perfect. There is no such thing as a bad child. There is only a bad parent. People don't know how to raise children. They don't even think about it. Most people who get married, and subsequently have children, never really consider the idea of how to bring up children. They just do it. They figure they'll figure it out. And you do figure it out. But, if you don't have a good idea of what is going on in their heads, of how they are developing, and what their view of the world is like, then you really can't relate to children properly. Neill believed in freedom without license. Which means that you let a child grow. You don't insist that the child learn table manners, don't insist that the child be mannerly or neat or learn this or learn that or eat this or eat that. As a matter of fact, it's becoming more well accepted that if a child is left to eat whatever he or she wants to eat, and is given a choice of all the proper foods, that the child will naturally get into the habit of eating what the body needs. And it's like that with everything. They will naturally eat what they need. They will learn what they want to learn and learn it well because they *want* to learn it. Their manners will be honest manners.

I've thought a lot about what would happen to Springsteen if something happened to me and I'm not sure. It's something that I really can't decide right now, because there's a possibility that Alan, Springsteen's father, will be coming back to town. He's been saying that for six months, but he may really do it. I still love him and if he wants to now accept some of the responsibility for Springsteen and comes to love her, and she comes to love him, then he might wind up raising her. It depends on how they get along, on how responsible he is, on what kind of terms we are on with each other. Something that I got straight with him from the very beginning, and we had a tough time when I got pregnant, was that I wanted us to be friends. I wanted us, no matter what, to always remain on a friendly basis. Because the child who sees its parents at odds with each other, even though maybe the parents try to hide it, always somehow knows and is affected by it. It has a lot to do with how secure they feel in the world. I think if they feel harmony between their mother and father, and consequently themselves, then they can feel harmony with the rest of the world a lot easier. If Alan acted the way he idealized, if he really did what he portends that he believes in, then we could live together very easily. But he doesn't. He has old-fashioned ideas about marriage.

I get scared, once in a while with the responsibility of rearing Springsteen by myself, but I know what I can do. I know that I don't *have* to have somebody else help me. I can do it myself, but help is nice.

Working at home also makes it easier. I like working for myself and I like my job. That makes it so that I am not yelling at Springsteen because the boss yelled at me. Nobody yells at me. I yell at Jasper, my dog. He's more convenient. I think working at home is definitely an advantage. I don't know very many people who wouldn't like to work at home. Or at least for themselves.

We live in a time when women are expressing their identities in more differentiated ways than in the past. One of the ways is being a single mother. It is easier now. If I had been born 40 years ago, I wouldn't have tried it. It would have been very difficult. Women are getting more freedom and respect, coming into their own. Springsteen's going to be the first woman President, maybe.

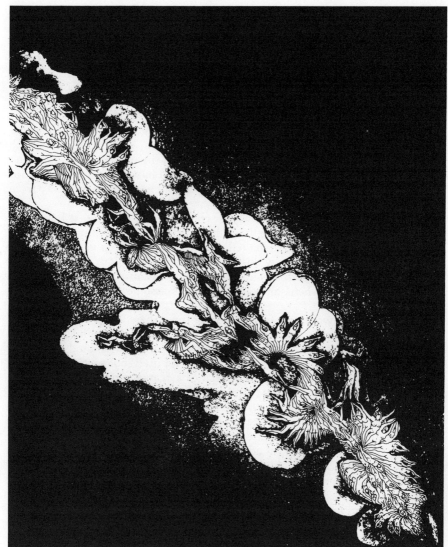

Arthur Okamura

UNCONSULTED FATHER

Stephen Anon, talking

Linda and I were living in the same commune in Northern California. We were there together, with and without Sarah, the woman I'm usually with, for a period of probably a year and a half. Linda and I got it off sexually. We were attracted to each other. I didn't have *any* idea that she wanted to have a child. I would have thought the opposite. I found out she was going to have my kid when she was about five months pregnant. I was living with Sarah when I found out. We all knew the situation. And it was either that the three of us would make a household with our (Sarah's and mine) other children or possibly regret the situation for the rest of our lives.

But, Linda wanted me to live with her, have this child there, and leave Sarah. She had a romantic notion that somehow she could prevail in the situation. When she found out that she couldn't prevail she decided to have the child anyway. I was in torment as to my relationship with the whole situation. I tried to be there when the child was born. But, that was impossible. Sarah and I had left the commune. It was really deep in the mountains and I didn't know what the birthdate was going to be. I could have lived with Linda for two months until she had the child. But that would have taken me away from my family with Sarah.

I think Linda chose me to father her child. I think that's what happened. But, you know you don't decide that a biological miracle is going to happen. She knew she was fertile and didn't tell me. I would have been very sensitive to the fact that she was. She knew this and just went ahead with the liaison. The result was the child, which I think she wanted because she did nothing to change the situation when she was pregnant. I assume that was part of the design.

After the child was born, for a period of time, there was a certain amount of what I felt was really manipulative communication. The child was, I suspected, being used as a reason for liaison between us. Because Linda did know I was very involved with children. I felt used. Not used, but sort of ripped off. I think Linda thought that I was somewhat of a dramatic or significant figure, and *that* was her way of relating to a significant figure, to have a child by him. I was an object. I *was* sexually attracted to her, but that didn't mean I wanted to partner a child with her. I'd say this was more of a power play on her side to have a relationship with me at the time. I know this from the name she gave the child. If she had told me, in advance, that she wanted to have a child with me. The first thing I would have asked myself is, what is my biological connectedness to this person? The second thought would be, if a child occurs as the result of this liaison what do I have to do with it afterwards? If the terms were that I wouldn't have anything to do with it afterwards, I would feel frozen. Because I'm involved with children. I was a child. To reproduce another human being on the basis of one person's choice in a situation is not fair human species politics. That's not fair politics, in my opinion. It's a form of fascism.

Another way to see this is when Linda got involved with another group of people and another man, and the child was not yet five years old, she gave it another name. To her the child was an object. She had originally given it a name that she thought would flatter, or be involved, with me. When she got involved in another group, it got another name. At that point the kid was a used car.

I haven't had any kind of opportunity to have a connection with this child because there were some legal reasons not to. It would have jeopardized the child, the mother, by my claiming paternity. It was a welfare situation. If I would have stepped forward and claimed the child I would have been responsible for all the money that had been paid out towards support and prosecution of the mother. It would have endangered everybody to do so. Linda would have gone to jail and the kid would have gone to a foster home.

It seems to me that if a single woman wants a child and wants a man's cooperation, then the man has to decide where his morality fits into it. I would suggest that *that* morality's on a human species level. We need each other to continue the species. Children that are born out of this situation, of a mother and a father, are in a desperate species bind. A child who never has either a maternal or paternal parent during his life is marked. There's no question about it. There's a dynamic between the sexes that's involved with the perpetuation of the species. We're not cars learning to make other cars. We're not machines learning to make other machines. We're human beings, part of the human species. It's the basis of our spirituality, our culture and ethics.

It doesn't have to be a nuclear family, it can be an extended family. In fact the isolated nuclear family of western industrial times is a very recent operation. It has never occurred before. I'm part of a heterosexual, parenting situation, but the children are encouraged to think of other people—who aren't necessarily blood related—as uncles and aunts. I feel good about that. I think they feel good about it.

The question comes up whether kids need the role model of a male or female in their life . . . that sounds like a machine to me. I don't think they need a role model of a male or the role model of a female. I think children need the dynamic, the interaction, between not only sexes but generations. If they don't have that, it's a big hole for them.

If a woman decides to have a child, be artificially inseminated with an anonymous donor, she will be conceiving an industrial age orphan. An orphan is an accepted concept. An orphan can happen naturally. A father or a mother could be killed, or both be killed, and an orphan could result. Orphans are well-known phenomena. It's just that with artificial insemination all the offspring are orphans. If somebody decides to have a child without a relationship to the father or the mother the result is a partial orphan, or maybe a complete orphan. I don't think that's necessarily bad, I'm just saying that that's the situation. If they're not aware of it, they're mad.

I could give you a beautiful example of an act of human compassion. A woman I know had a child and wasn't feeling that she was capable of raising it. So, rather than leaving it on somebody's doorstep or giving it up to an orphanage, she asked a good friend of hers whether or not she could take the child. And the friend consulted with the man she lived with and they agreed that they could. These people are terrific parents. That child today not only knows that somebody else is his mother, but also feels that these people have become his parents. And he's a splendid child. That's an example of orphanage occurring, without any male/female donor rigamarole. Orphanage will go on forever. It is a natural part of life. How it's handled is an important thing. If you eliminate a child's connection to human species continuity by your deliberation, they are going to catch you at it. They are going to *find* out and they are going to demand that that be filled in. I think that's a human species phenomenon. Kids aren't kids, they are people. They're not objects, they're people. They are not machines, they are the same as the parents. They have as much dimension and profundity and maybe more.

Single parents, who exclude the other sex, may desire to remake society.

What they're remaking society into is deeper than they may be aware. If you have a seriously damaged childhood, there is no reason to believe that an undamaged adult exists. You can feel that whatever happens to any child, as long as it's born, is alright. Because you made it. But, that point of view seems to me to be particularly lacking in compassion or depth. Because regardless of how hurt you might be, it's like being a handicapped person. Does a handicapped person want to have a child that's blind because they're blind? If a handicapped person has a child that can see as a seeing child, then what do they do? Go beserk? Go crazy? Have they failed because they had a seeing child? People need to think about how much of their own past they are unconsciously projecting and recapitulating onto their kids.

In Linda's case, when she had an opportunity, seven years after the child was born, she left it. I think that the child was being used as an object and that at a certain point the objectness of the child became unuseful to her. Fortunately, the child knows where I live. Without trying to sound romantic, I think it's inevitable she'll come to live with us. Sarah and our kids are prepared for it. Everybody's aware of the situation and hoping for an early reunion.

WORKING WITH SINGLE MOTHERS

Barbara O'Hara, talking

I was a birth assistant and counselor to expectant mothers for two and a half years. During that time I assisted about 170 births, mainly in Berkeley. A little more than half of these births involved single mothers. Most of the women were between twenty and twenty five or over forty. There was hardly anybody in the middle range. Single women in the middle range seem to be more intent on pursuing careers, working on themselves, or being involved in all sorts of social activities. Once single women reach forty, a number of them seem to decide that they want to have a child. Their focus and energy seems to change. They give up trying to find the perfect relationship and just decide to have a child.

I worked in one clinic for several years that did, among other things, artificial insemination. The women who wanted this done were usually older and had read a great deal about it. They were medically very knowledgeable, knew where the sperm banks were, knew even how recent the sperm was and all that. Some of these women didn't want to know anything about the donor except his race. They were very clear about that. Both black and white women wanted donors from their own race. Others were women who hadn't been able to find a man willing to father a child. These were women who had put the question to several men they had been sleeping with, without success. The men just freaked and said, "What do you mean, have a kid?" Finally, there were gay women, who didn't relate to men at all, who wanted to have a child in their lives.

The doctor I worked with, who did the actual artificial insemination, did a lot of what we called psyche screening. We wanted to find out where these women were really at, what their unspoken dialogue was. A lot of these women were lonely and wanted a child. It was real clear that what they wanted was this person loving them, this child, and getting to love it back. They wanted *unqualified* love, as they put it, from a child.

What we found over time was that many of these women questioned their choice of having been artificially inseminated. They had many doubts. They would come into the clinic for a check-up and hear some of the other expectant mothers—who hadn't been artificially inseminated—say things like, "I think my kid might have blue eyes because George has," etc. At around six or seven months, when the inseminated women could feel the baby move, they would want to know more about the sperm thing. But all we could do was laugh about it, and say there was no way we could trace that back for them. That had been clear at the start. But now they were pregnant and had to deal with being pregnant from an unknown person. These women now realized their kid was going to be asking all these incredibly important, emotionally packed, questions. We had cautioned them at the start—when they were first considering being artificially inseminated—to go home and think about this. We said to them, "how are you going to handle it when your little boy or girl says, 'was my daddy tall?' What are you going to say? 'Gee, I don't know, he's from a tube,' he was from test tube three.' What are you going to say? Think about this, this is a serious decision here." Well, some of the women decided to do it anyway. They felt that they could handle it. But our experience was that they couldn't. We finally stopped doing artificial insemination because the end result was always a crisis and very heavy for the woman. She would feel very sad after birthing, on top of the normal postpartum depression which women experience. It's hard enough not having anyone to share a child with, let alone not being able to share a child's history and characteristics with it. Roots are a thing that bother people. Kids really want to know their roots. We all got very depressed by what followed upon artificial insemination, so we stopped doing it, after awhile. The doctor said, "Hey, I don't want to do this anymore, I don't like this. It's awfully strange."

The younger single women, between twenty and twenty five, who decided to get pregnant and have a kid were another whole set. When we asked them why they decided to have a child, they would giggle and say, "I don't know, I just think I would like to have a child now. I've got my energy and passion up for it." They thought it would be a beautiful and wonderful experience. They felt that they could do it in Berkeley. They felt they had enough support systems happening here. They mentioned things like recycling centers for children's clothes and toys, food stamps, a lot of women's support groups, single parent's groups and so forth. This was the time they consciously chose to have a kid because they wouldn't be doing it essentially alone —even if they didn't have a man. They could get help from the community. They knew a lot of other single mothers. If the women were under 20, they generally came to us because they wanted an abortion. Most of this group didn't want to have a child, yet.

We would never work with a woman who had tricked a man into fathering a child or who would come in and say, "I'm pregnant and it's this man, but I don't want him to know." We didn't want to be put in a situation where a man would call us up and ask if a particular woman was coming for treatment and be expected to lie for the woman. I told them if a father called up I would tell him the truth. We didn't want some man to be coming in nine months later, during delivery, screaming at us or her because we had hid the facts from him.

Most of the women who had consciously decided to have a child without a man were in their forties. They were pretty much into wanting to be alone. They had *had* it with men. They were bitter or pissed off from before. A lot of them had had one child already and dealt with a man where it didn't work out. And yet, they loved having kids. So they were sleeping around and had relationships with two or three men. They had relationships and friendships with the men, but didn't want to live with them. Then they would put it to the men, "How would you feel if I got pregnant with you?" The men would say, "What do you want from me?" And the woman would say, "Nothing." That tended to freak the men out more than if the woman had said, "I want some financial support, a hundred a month." But *no* input was difficult for the men to deal with. What we found was that the men these women tended to drift toward would get hysterical and say, "Okay, fuck you. Here's a kid, you want a kid, but don't bother me ever again." These women would go for that kind of man, rather than the one who would give them support. They'd pick the man who was pissed off, figuring he'd never want to see the kid again. We'd sit around, when the women explained this kind of situation to us, and say, "Okay, okay, we'll see, we'll see. But leave yourself open, leave the door open in case either of you change your mind."

The women had it all verbally figured out that they would say the father, "is this man who lives in town, but he's not interested in having children." They wouldn't badmouth the man to the child, they just wanted the child to be totally theirs. They were totally into power tripping.

But what would happen, as they got more and more pregnant, is that they would have fantasies. The power thing became less important. Their bodies would go into some other state. They would say, "Power? Oh, I can't be bothered. I have to go to the market now and get my vitamins, I have to get my green vegetables. I can't deal with power right now." They became very drifty once the sleepiness and nausea of the first three months of pregnancy passed. We'd all sit around in a circle every morning, all the nurses, the midwives, the practitioners, the kinesiologists and fill out each woman's chart who was in the stage of being alone. Yet, we all knew that she wasn't really alone. The man was in town, in the garage around the corner, or he would drop by after work and they would fight. For the first three months they would be fighting and she would say, "It was your choice, I told you I just wanted you to get me pregnant, don't bother me about it now." And the guy would say, "Well, I just want to make it clear that I'm not responsible." For the first three months there's this dialogue about "I want to make it clear, I'm not responsible," and her saying, "Of course not, it's the way I want it. Stay out of my face, don't come back." Then, when the woman would start to "show," something would happen, would click with the man, and he would come by and he'd say, "Do you have enough food?" That's how it would start. "Do you have enough food? Are you eating? Are you drugging? This is my kid." And she'd say, "What do you mean this is your kid?" and this whole new dialogue would start. He'd say, "Where are you going for care? Not that I care, but where are you going?"

Then pretty soon they'd both come in. Part of our package deal with single mothers was that they had to come for the therapy sessions once a week, so we could check them out. Because we wanted to know, at any point, if they were not going to be psychologically ready for a home delivery. We wanted to know if they were unconsciously planning to have a bad delivery. So we watched them very carefully to make sure they were clear. We did massage to relax them and a lot of body-strengthening stuff. Pretty soon the woman would say, "Get ready, Mark's coming in. He's being a real asshole. He's coming in next week, just being nosy and I'm bringing him." And then we'd all start laughing, "Oh, you're bringing him in, oh you're bringing him in, you want to think about that? You want to hear what you just said, you're bringing the asshole in." And the smile would just slowly come to their faces. "Okay, so he's interfering a little, so I'm letting him in. But you did say, "Keep the door open.'" And we'd all say, "That's right, we did. He might want to share in this. We'll see how it works out." And, little by little, it turned out that 70% of these women, who started out choosing the man to have a child with, but not be responsible for or want to have any ties with, 70% of these men were there for the birth and were tied to that child as much as the women were. They were not pulled in kicking and screaming. They were there, they wanted to be part of it. So the whole idea that men could accept just being studs didn't wash. It was almost non-existent. It became a thing of "There's a part of me, I have a child out there." It became an emotional thing that nobody was able to stop, which was really nice to see. This was a great support for both people.

I now go around Berkeley and see many of these couples together. It's a real kick, because I like to tease them. I say, "Oh yeah, isn't that the guy that you said was never going to be around?" The woman's response is often to say, "We're going to have another baby. But don't worry, he's going to be there the whole time." And the guy will say, "Bullshit, I was there the whole time anyway, the first time. But this time, I'm going to be there from the very beginning. You know, I'm not just a servicer." And it's turned out real well, Berkeley being so loose. It was an eye-opening experience for me and for a lot of the women of my age that were midwives and practitioners. Because a lot of us, ourselves, had gone through bitter experiences. We found out you can't put down a man who comes forward and says, "I'm concerned, I want to be part of this." You start thinking, "Well, gee, maybe not all men are awful." All of us softened.

Marriage didn't enter into it so much, the living together thing happened. Being a family happened. Watching them walk around Gilman Street at the

Deli, it was a joy, seeing them come up with this child. It was a greater high than the women who were alone after artificial insemination and who had that heavy impact to deal with that I saw. I mean, I would never choose artificial insemination, it's an insane way. And it doesn't work emotionally to have an estranged man live in town with you and have him know you're carrying his baby, either. It very seldom worked unless the man was a merchant marine who was here like for one day and never was going to be seen again. But the women I assisted never picked men like that. It was phenomenal, there was a reason they picked a certain man to have a child with. When it got bottom-line time to get pregnant, they chose the guy that they knew in their heart of hearts would be around. Women in fact do pick men who on some level they feel they could have a family with.

Even if I was going to lie to myself and say that some man is "an easy mark because he doesn't want kids. He'll never bother me." I would never choose a man who would never want kids. Because I would want my kid to have some love from a male point of view and would pick some man who is soft enough to be touched like that. Not every man or woman is that soft, to let a kid in. And men's complaint, when they came into the therapy sessions really pissed off half the time, was to say, "There's no way you can tell me she didn't *know* that I was going to be involved, there's no way you can tell me that she didn't pick me on purpose." And it would come out towards the end as a very tearful revelation to her, as well, that she knew he would love to have his own child. Because he really had to let her know that it was not quite fair to unconsciously go out there and pick a man and do that to him. Not fair to him, or to the child, or to her. It's not done on a level as an adult to an adult, it's not the way to start off. Because if I were a man, I'd feel for a good 20 years a little bit manipulated and ripped off.

So here we are, a whole generation of ripped off people, trying to relate. Men walking around saying, "how do you get laid in Berkeley?" Women walking around saying, "how do you get a man in Berkeley?" And my answer generally is, I guess, be real direct. Women shouldn't have these unconscious hidden agendas going on. Men really have a lot of anger about being pushed away, and then seeing the child. The men that really felt ripped off were the ones where the woman would leave the area. They would work with us for awhile and then disappear. They would go home to their families, whatever, to have the child and leave the man hanging with this tie clear across the United States. He never would know how it came out. The doctor I worked for would say to these men, "You have to come back, you need help, you need a support system, you feel emotionally ripped off and you have been." He started men's groups and men's support groups. So the men wouldn't go the next 15 years feeling too afraid to have another child because they felt tricked the first time.

• • • • •

The third group of single women I worked with were the 20 year olds who felt that they could survive here, and that it was time to have a kid. They would pick a man and have this rational, intellectual agreement. They would have a written contract, specifying that the man would not interfere in the raising of the child. This didn't work out well because of the emotional thing that happens when you see a child that's your own. Many of the men wanted more input. They didn't want the woman sleeping around or other men coming around. They wanted half-time parenthood. They didn't want to be a weekend father. They didn't want to be a one day a week father.

They said things like, "I want that kid to know he's mine. I want that child to have my name, I don't want you dragging him from here to Europe for the rest of his life. I want some input here, legally." They realized they made a big mistake by thinking they could just casually come in and out of a child's life, and casually come in and out of this woman's life. The burn out for that was real fast. It happened two ways; the men would go to court and they would be committed, their guts were really committed to having a part of the child's life. Or it would get real messy with a lot of quarreling and bickering, and all three people being pretty much emotionally disrupted for a good fifteen years.

A child has a magical effect on people that no one's prepared for until the child's in front of you. You see part of you in that child. It brings out all kinds of unexpected and uncontrollable emotions. Passions, anger and frustrations of being ripped off. And everything else seems to fall away. The job, money, security, drugs, all the rest seems like a pain in the ass compared to the reality of the child. The man wants to be able to have some input and asks, "How come this dumb girl's walking around saying 'Heh, you know, sure you can drop by anytime you want.'" And think of the confusion on the part of a child when it asks, "Who's my daddy, where is he?" Look at the number of scattered children who are twelve years old on the street, not having any roots themselves. I know this man who has a daughter that he casually gave birth to 13 years ago. She just knocks him on his ass left and right. She buses down here from Humboldt every weekend, she runs away from home, comes down and buzzes him. This man is going through his late 30's, early 40's throes of, "What's it all about, Alfie." He's got this 13 year old coming around saying, "Hey, I'm smoking dope, I'm having sex, I'm doing drugs, what do you care? Where the hell were you for 13 years?" He's stormed in and out of her life for 13 years. He casually had this affair with this woman, she did the same thing, and here's the product of that. He didn't know, she didn't know, their child has no roots. She's a very strange child. She's got a lot of anger towards women and men. And she's almost wild beyond belief. She has pure energy and no roots and no ethics and no morals and no adult that can say to her, "We were there when you needed us, we've given you this and this and this." Bullshit, neither of them were there, they were there casually.

But more often, despite what the original agreements were, I've seen people come together. Because of the man totally realizing upon seeing the child, and because the woman is not prepared for the man's response. You'd think they would know each other to have this child. No one's really prepared for that response that a child gives you. And men tend to surprise women with it by their being there. The women walk around saying, "They're full of crap, they don't want you, they don't know what's real," and then the man's saying, "I'm telling you I'm here, I want to be here, I want to love you, I want to love the kid, we'll try it out, we'll make it work."

I myself am going through a divorce and I have a child. Her dad is a wonderful father. He tells me at every chance, "I'll not be a weekend father,

I will not be a one-day-a-week dad, I'm getting her exactly 3½ days a week." I really respect the passion that he feels.

Even though people don't stay together it doesn't matter to the kid as long as they have that connection with both parents. As long as they feel that there is someone, that they've got two people out there that will really take care of them, protect them, always be there. Men and women have to be clear that their relationships with each other are separate from their relationships with their kids. People have to get it clear from the very beginning. That young child has to be reassured, no matter how we change, you and I as people, no matter how many other relationships we have, that it has two parents. A father has the right to say, "I want constant input to that kid, direct input. I don't want it screened by you. I want to be able to directly call him and let him know, 'I'm your father, if you ever need me in the middle of the night, if you ever need me at Christmas when you feel alone and all the other kids' daddies are there, call me!" Step-dads don't always make it and there's no such thing as a replacement. Once kids know that their father, their mother, is alive, that's it.

Kids are like the walking wounded if a parent is missing. There's a part of them missing and they want to know where it is all the time. There's a search which is endless at night that kids go through. Of, "If he loved me, he wouldn't have left, I don't care what my mother says. If she loved me she wouldn't have left him, I don't care how great this new woman is. I know I have a different daddy, I know I have a different mommy." And that pain can be eased for the child, in their heart of hearts, if they have direct input from their parents, constantly through their twenties. That child can grow up and say, "It's alright, mom, my father was there. Jack's nice but my father was there and I love him. He gave birth to me, he's my father, he's my blood." And there's no replacement for that.

We're a rootless country right now, these scattered kids are everywhere. I have a friend who was one of twelve kids. He's got eleven brothers and sisters and says, "We don't know who our fathers are, and it made us all slightly crazy. I'm going to get married and turn this thing around." He's 35 now. He wants to get married, have a kid. He tells me, "Even if that relationship doesn't last, the marriage doesn't work, I want to be able to say to my kid, 'I was there when you were born, I am here now and I am going to be here, I am your father, hello. I don't care if you're in Africa, I don't care if you're in Sweden, you'll come back and we're connected.'" And what that kid will respect and love is that the father or the mother chased them down for twenty years to remind them, "if ever you want to come here, I'm here. If ever you run away, would you please run away here." My mother had three marriages, for instance, and I would have given anything if my father had ever come back once and said, "When you run away, run away here." And I've known a lot of my friends to feel that, too. I'm in my forties and in my generation the biggest sadness was not knowing where your father was.

But, now it's turning around, mothers are splitting. So in the next generation we're going to see all these kids walking around saying with bitterness and loneliness, "Where's my mother?" Because the men are standing still more and saying, "I'll take them."

The more vocal feminist position that still characterizes men as being insensitive and not wanting to really be responsible is erroneous. I never hear any of these women come out and acknowledge the high percentage of men that are becoming full-time parents. I see these men. I know them myelf. They are better parents because the women are breaking. They're backing away and saying, "I want to live, I missed my childhood." Things men were saying when they took off last generation. The women are splitting and the men are not falling apart. Men are saying, "I think I can do this. I'll stand here and try it. I'm not going to dump them. I'm going to raise these three kids." And they go into the same depression, tiredness, exhaustion, that the single mothers used to go through. And there's a support system out there. There's a Single Fathers support group. My ex-husband, for instance, belongs to Parents Without Partners, PWP he calls it. There's an incredible amount of single fathers who have full custody of kids. It's not investigated, it's not in print, we don't know how many men are doing it.

In Berkeley, maybe half of the men are raising their kids. If not alone, at least half-time. They're doing it real well. They've got their briefcase and their tie on, or they got their mechanics clothes on, and they are going home to feed their kids—the same as I'm going to go home to my kid from a full day's work. And they're getting off on it. They like it. It seems to really have grounded them. The courts are changing, too. When I went for a divorce, for instance, my lawyer said, "Don't think it's so easy here, to walk in and get your kid. It used to be five years ago, but not anymore." Judges are more conscious. If a man shows up in court and says, "I want my kid, at least part-time," he might get him more than part-time. Because the judges see these people come into court and they see the mothers trashed out and the fathers saying, "She can't handle it, I want that kid, or those five kids." And the judge says, "You got them." California is no longer a "mother's state." They used to call it the Great Mother law that women were naturally entitled to their children. It's just not so. It just depends on where the parents' individual values, morals, needs, and capabilities are at that time in their life.

Some women come back, after they've split, and get blown away by the fact that the kids have done so well, and the man's done so well, and there might not be a place for her. That's why I always tell people, "Don't just split, let these kids know that you're there for them. Because they might not want you to come home. They might not want to come to your home. Even if you can't cope, at least let them know that you're not able to cope at this moment, but you might be back and you love them very dearly."

The feminist push about most men still not being there is just not true. The rap doesn't work. Okay, we have jobs, we got some more power and we're equal. But when you get that, you have to look back at the man, and if the man's not screaming and yelling about you being equal, then what are you going to fight about next? There isn't anything else. You are equal. And we are. And that's why they can't bitch anymore. It's changing and we're going to have to go on from that plateau of "Okay, so we're equal."

Kristin Wetterhahn

CITY/COUNTRY MINERS

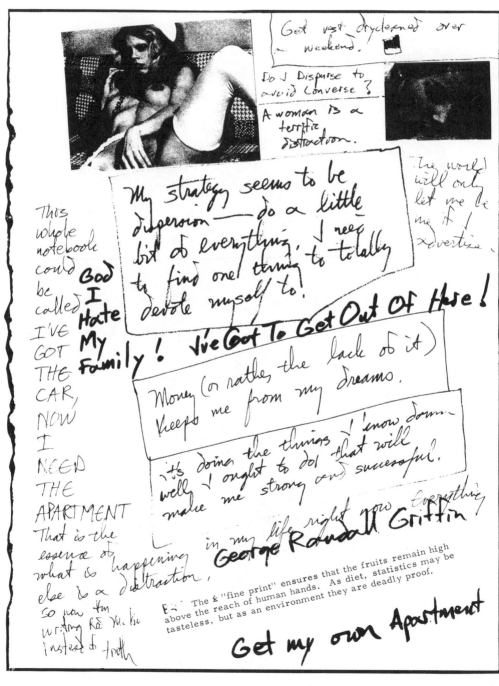

George Griffin

Break Out

You're the perfect hero . . .
if they put you in heaven
you'd find a way to
        break out.

    My room is starting to look like a picture in a magazine
    Telephones are growing videos, cars hate Iran, my typewriter
        used to be my lover

    She walked into my magazine, I took her for a spin in my hate,
        typed out a sexual transference poem, she called it love
            on the video

Written by Agony

if it comes to you easy
it'll leave you rough,
if it comes to you hard
it'll never leave you

—George Randall Griffin

what do you say
to your former old lady
when she and your daughter
are asleep on your bed
after a party
on christmas
late at nite
when you got up two hours ago
and fixed dinner
and the birds are done
and you have to get up
and go to work tomorrow
and you're not even sure
if you're hungry?

      ego! ego! ego!
the shrill cry of a lonely nite bird on the wing
      ego! ego! ego!
leading me in a dance of living death of no rebirth
      ego! ego! ego!

—Tom Hile

Alison Palmer

MORE LOOSE LEAVES FROM THE LITTLE BLACK BOOK
Jennifer Stone

Fall 1967 (The Creature from the Black Lagoon is your Father)

What was it he said to me last night; "Beloved Pussy," he said, "you've got the baby and I've got the ball." He goes in for the D.H. Lawrence stuff, the blood coming out of the earth bit, the ships that clash in the night theory of sexual encounter. He doesn't mix up lust with love the way I do. It's the same old Yang-Yin mess. Like when I want to get in touch with my primal self, I sit near the entrance to a cave on the beach. I wait for the sea to give up her dead. I wait for the tide to turn and then I read the seaweed like tea leaves in a cup. He, on the other hand, wants to make it with a dolphin.

Now I'm not saying all men are obsessed with the erotic or that all women are stuck with romanticism. *Au contraire.* Some of the men I know are so romantic they fuck only goddesses, or at least princesses. This cuts down their sexual outlets, raises the crime rate, costs women a fortune in cosmetics, time, energy and self-respect, but even I used to think it was fun, once. I blame Greta Garbo. Then there is another kind of male romantic who fucks only whores and worships mother-madonnas or daughter-virgins but everyone knows all that stuff nowadays.

Consider the male romantic when faced with pregnancy. Faced is the wrong word. When I was pregnant, my husband said he was too much in awe of my condition to fuck me, which is a damn lie, he thought I was ugly, but at the time I figured he was a gentleman to lie about that because of course I thought I was ugly too. Man as mirror and so on. However, my hormones ran amuck when I was pregnant. I was painfully horny. One night I went to a swanky bar in San Francisco. I picked up a kinky character who really dug pregnant women because he was so afraid of fatherhood this was the only way he could be absolutely safe. He even wanted to see the stretch marks.

There was a pink douche bag in his bathroom. I asked him if he always left it sitting on the back of the toilet like that. Well, he said, nobody wants a smelly woman. That seemed reasonable enough. Pregnant is bad but indelicate is worse. I went to the kitchen and got a bottle of coke out of the refrigerator and filled the douche bag with it. Then I hung the bag over the shower bar. Before getting in the shower I thought about it some more. I wanted to be fair. I wrapped myself in a big bath towel and went into the living room. He was setting up for our orgy. Psychedelic lights and music and all. I think he wanted to make a movie. For all I know he did. I told him about the coke douche I'd fixed up. It didn't get through to him right away, so I asked him what he was drinking and he handed me a bottle of Jack Daniels. Good idea, I said, returning to the bathroom, just what it needs.

October 1967

(symposium: in ancient Greece, a
drinking together, with
music and singing)

This one is different. My children are gone for the week. We are alone and in love. We scrub each other's backs and feed each other strawberries and lie on the floor and drink wine. Gratified desire, satiety, the works. We walk out to a cafe-bar to pity the loveless world. We are so knowing and so self-satisfied, everyone is drawn to us. We are holy and they want to touch us. We don't speak often, only to confirm each other. The shared laugh of seeing through everyone we meet; to understand is to forgive . . . we forgive everyone. And when we are alone again the gong sounds like it does in the beginning of a J. Arthur Rank film and we are drowned in each other and if it weren't real it would make a swell movie. It's never been anything like this except once when I took acid but the man I was with then was far away. This time he's more me than I am and we have been here together many lifetimes before and we are drinking each other alive. This is the moment to die. Nothing can get better. Nothing does.

November 1967

Sam and Simon and I are walking around Jewel Lake in Tilden Park. Simon talks a lot. He asks if we remember the winter the lake froze and the ducks slid around on their tails, and do salamanders drown, and if Christ was a carpenter did He make coffins and was He *really* a Jew or a red herring. Some sweet elderly women in tennis shoes and sailor hats ask Simon if he knows what kind of bird is the little brown bird on the water, and he tells them it's a baby duck and even baby ducks have trouble landing on the ice.

Sam runs into Rebecca. Rebecca is his sometimes, well if he isn't busy, and there isn't a ball game, his sometimes-go-and-see-her girl friend. She has a six-pack and Sam follows her down the trail and Simon and I sit by the water. Simon talks on about Sam and his friends and male bonding and silver-backed gorillas leading the pack and I ask him if there isn't more to adolescence than sex and violence and he says of course not.

Sam comes back alone and says he wants to go and see *Godfather II* and Simon has to sit down front because he talks too much. I remember my mother used to make me sit down in front.

A brown mallard hen is running along the edge of the lake. Her back is raw and several larger ducks are pecking her in a determined, impassive way. Simon grabs for the hen who seems more frightened of him than of her killers, and a band of black children appear on the path and chase the pecking ducks into the water, throwing stones after them to drive them across the lake. They yell after the ducks, "Honky motherfuckers!" Sam and Simon squeal with laughter and join them, throwing stones and screaming "Nigger cocksuckers!" They all roll on the ground together and snort with laughter.

Over pizza, Simon considers. He shakes his head and says he's no honky. A honky, he tells me, is a white male adult who drives around Harlem honking for a whore because he is too chickenshit to get out of his car and walk around asking for one. Sam says that makes a lot of sense.

Sam says the reason *Godfather* is more tragic than a western is because they used to be Romans once, they're older barbarians. Simon says no, it's the musical score which is a dirge and the movie's a funeral and that's why it's fun.

We have chocolate sundaes. Simon mutters to himself. I ask what is bothering him. He says who in the hell *was* the Red Baron really. He is looking at a large stuffed Snoopy doll in the ice cream parlor. I tell him I wasn't around during the First World War. Neither was I, he says, and neither were the Romans. First thing in the morning Simon is on the phone to the public library. He asks for the Red Baron's last name and he writes it down. The Red Baron, the World War I flying ace, is named Manfred Albecht Richtosen. As they used to say in the newsreels, time marches on.

Winter 1968

I met a man in a radical therapy group. He gave the impression of warmth and civilization. I wasted the usual hours making myself attractive before seeing him again. I wanted to be sensual and civilized. The first time we were together he read me a story about a venetian mermaid: "The Copulating Mermaid of Venice, California." It's about two drunks who steal a cadaver, discover it to be the body of a beautiful blonde woman. They rape her remains and swim her out to sea. This was the first stage of foreplay. I dropped a cup of tea in his lap to make sure he could still feel. Then I read him the autopsy of Marilyn Monroe. It left him cold.

Winter 1970

"And unto them was given power,
As scorpions of the earth have power."

When the neighborhood rapist appeared in my bedroom at three o'clock one morning, I was sure it had happened before. I recognized the feeling as if it were death. I knew all along it was coming. *Deja vu.* Black, with a hat on. As gently as an ancient, I asked him what was the matter. He told me. I comforted him with all my skills. He assured me he wouldn't hurt my children if I did what he told me. He had been through the house and seen the boys' blond heads on their pillows. What he could see of my room in the dark made him think I was Asian. I could have been. I could have been anything. Geisha woman or goddess.

What must be understood is that he believed me.

A man found a ladder in the basement and broke into my home through the bathroom window. He threatened the lives of my children and would not let me move or turn on a light. And under these circumstances I convinced him he'd made me happy. He promised not to frighten the children. He was lavish with his compliments. He said I had a snatch like a sixteen year old. All his misery poured forth. He talked about the lousy treatment he'd gotten from women. All they needed was a rod in the right place, he said, then they knew who they were. I consoled him.

He began to give me advice. He told me how careless it was to come home alone late at night. A carload of friends had driven me to my door about midnight. I'd forgotten my key and gone around to the back, making a lot of noise and calling attention to myself. The porch light was on, he said, so he could see I was wearing a mini-skirt. He told me he could break into any house in the neighborhood and what was I doing living alone in a place like this, was I on welfare? I told him I was a high school English teacher and he was very impressed. He told me I must learn to be careful because this whole neighborhood was full of real bad men and they could be watching my house.

Finally he got out of the bed and said goodbye with a sort of wistful sneer. He would go out the back way so as not to wake the children. After I heard the door close I tried to get up. I moved to the end of the bed. Some time later I heard a dog howling somewhere. I tried to get up again. I walked along the wall until I came to the back door. I locked it. I got to the living room and lay down on the sofa. After a while I could hear birds outside. I found the children buried in their sleeping bags and I unzipped the bags to make sure they had enough air.

It's been a long time now. I sleep in my clothes. I lock my children in their rooms and lock myself in mine. Sometimes I sleep on the sofa next to the phone. The worst thing is I don't know what he looks like. Black and high cheekbones but that's about all. Every time I see a black man with his profile, I have to wonder.

Finally I move. Retrench in a safe neighborhood. It's my own fault. I thought I was a white man. I thought I could live as I liked and the world would love me for being myself. Zen slap. My encounter with the facts. I've got to forget it. It happens all the time.

The nightmare always comes back. It's always the same and it's always different. Sometimes he comes through the walls; sometimes he just drops by with friends and I suddenly know it's him and he was there all the time without my seeing and when I realize this I can't breathe and I am smothered awake. It's the feeling of suffocation from fear, it's the *I can't scream* dream.

When the police took me to the hospital I was afraid to give my name because my older sister works in the medical records office there and if she found out what had happened I'd never live it down.

Summer 1974

> . . . I carve a limestone fossil of a god
> laughing at my lover.

It happens every seven years. Transfiguration: a new self. I was thirty-three when I met him. Forty years old now and he still calls. I don't hate him, I'm just past caring. I mean seven years of one night stands with one man. I can't go on treating him like a sex object when I've forgotten where I met him. We said everything we had to say in less than a week. He's old now and cranky and paranoid and something of a bore. Rather like me. Only I don't expect undivided attention anymore. Oh, I demand it but I don't really expect it. He wants me to listen to his every word and he had nothing to say when he started. He roars a lot. I have no idea what about, but the noise impressed me for quite a while. Then it irritated me. Then I thought it was sort of sweet. And now I'm bored. Sound asleep if you want to know. If things go on this way it looks as if I might turn celibate. I'd get lots more work done. Balzac said a night in bed, in bed with a woman I mean, cost him any number of pages, I think he said eighteen. Balzac thought of it as a physical drain. For *me*, it's a brain drain. When I'm in love I think all the time. I analyze every word, every gesture. I pick and dig for evidence of criticism or censure or even for approval. Scratch, scratch, scratch. He quit years ago. He hears nothing. Oh, he mellows a little if I lay it on with a trowel, if I act ecstatic and gratified as hell. He basks in my afterglow. But if I'm cross or gloomy he just ignores me psychologically. Never biologically. Result: I am impassible.

Yes, it may be true the body renews itself every seven years but brain tissue is permanent. It modifies but it doesn't renew. I only get one soul. I know this is true because when I run into people I haven't seen for twenty years I can get away unrecognized if I don't speak. If I open my mouth they can hear the same old nervous system. It's my father's mother's voice I think, more than the rest. In any case, it never changes.

Perhaps it's time to quit. Time to quit talking, time to quit fucking. Ghandi quit fucking when he was thirty-seven. Seminal fluid to pituitary gland; energy turned spiritual, that sort of thing. All that heat going to the brain. For once I could get out of a male head and into my own. It's very contemporary, living alone and doing one's own thing. One's own thing. Beats getting all dressed up and spending an evening being gorgeous just for the sake of a few orgasmic moments or even for a few non-orgasmic moments. I'll give it a whirl. I'll even give it a title: Neo-Narcissism.

October 1975

My old friend Julia insists we go to our twentieth college reunion. She has a sadistic streak. So-and-so, she says, is dying of cancer. I don't remember so-and-so and this doesn't seem like the right reason for a class reunion but I've forgotten how the bourgeoisie live and I decide to go and see. Julia and I went to a girl's college. Private, arty, and expensive.

We drive to the Piedmont hills. I've pinned myself together in my poet's drapery; black shawls and hat and red feathers. Julia is chic and svelt and she is hurt by my feathers. She smokes furiously, with a Bette Davis flourish. When she gets all dressed up, she starts acting like a female impersonator.

We arrive at the Alpine Circle in Piedmont, the home of Dr. and Mrs. Tong. I cannot remember the name of Mrs. Tong who was in my graduating class in 1955 so I have to call her Mrs. Tong. Haute bourgeois: four bathrooms, indoor pool, several sunken T.V. pits, rumpus room, the works.

Most surprising is the age of the people. There are rooms full of middle-aged and even old persons. The best-looking man in the room is pouring the drinks. Julia and I had promised ourselves to stay sober but I can't handle reality. After several bourbons, the man pouring the drinks remembers he went out with me the night he met his wife. This reduces him to maudlin tears. After that it's all a celtic blur. There's a woman who still wears her hair in a style that doesn't look like a hat. She once stole a man I don't remember really wanting but I didn't want him stolen. I ask her if she remembers what a wicked femme fatale she was and so promiscuous and her father a clergyman and all. She has had a lot to drink and she tells me that was only because she was impotent as a young woman and had no orgasms until she was thirty-seven. None, not even of any kind. She says I've done better because I have a space between my teeth and that means I'm sensual. It's a rumor that got started in Johannesburg. About the teeth, not about me. She asks me about the dates of my first orgasms. I tell her I can't remember and she thinks I'm showing off. I tell her I remember stopping off at a gas station on the way home from school when I was ten. What a shame, she says. Not shame, privacy is the issue, I explain. She cries a little because her daughter is a poet like me and no one loves her daughter because her daughter has a size 38D cup. She means that other *girls* don't love her daughter which is just as well I think.

Julia is as tight as I am. She is deep in conversation with a woman who is saying to her that if their friendship is to be revived it will have to be on the basis of mutual respect this time. Anyone who knows my friend Julia knows she suffers from terminal superiority (it's her animus, for the Jungians) and anyone who truly loves her must put up with it. She knows what's right. She's consumed with conviction. I launch a drunken defense. Julia looks as if she might throw me in the pool. I tell her my threshold for insult is too high for her. She announces it's time to get what's-her-face (me) home and she drags me out trailing feathers, both of us dead drunk. We talk on the same bumble-booze-brainless wave length. Great friendship in our hearts of course while she rages about the Tongs and the middle classes. She seethes with socialism and red righteousness and I try to contribute, Marx-wise. Then she tells me what a fool I am to love men who don't give me money. I ask her how old she was when she had her first orgasm and she takes me to the Pup Hut and orders coffee.

In the morning I discover I've lost my little black notebook. It's in one of the Tongs' bathrooms. I call and Mrs. Tong is Kim and the notebook is in her desk and she says nothing. Perhaps she didn't read it. All during the evening I'd kept hiding in a bathroom and writing notes about people. There was the chunky set in the living room sitting in a circle around the food. I dipped in and out of the room to throw a fish into the conversation and get back the bait. A red-faced bellicose male barked at me, "Are you some kind of feminist!?"

"Isn't everyone." I asked.

"You must have known my wife," he hollared.

I might have known his wife once, before the flood. I might even have known Kim, or the woman without any orgasms, or the man who met his wife, but it's too long ago now and those people died.

"Never, I assure you, never in the biblical sense." I kissed the red-faced man and he grinned and kissed me back.

End of September, 1975

Noon at the Bay Area Rapid Transit station. I sit on the bench and look toward the Berkeley hills. The hills are green. Two pigeons fly down from the sky. They soar through the open roof of the station in a great arc, curving into the glass windows leading to the green hills. I expect the glass to shatter but there is only a dull thud. One bird falls, limps, lurches, and drops quietly, head falling softly to one side. The other writhes in a furious flap of wings, her feathers flying, twisting in agony as if her back were broken.

A woman sitting on my right says well how stupid can you get. She says she thought pigeons were smarter than that. In a final desperate lurch, the broken bird falls off the platform across from us, landing on the train rails.

A woman sitting on my left, a woman with shellacked hair and eyelashes glued tight, says well it was probably a male chasing a female and serves them both right.

FABLES *Jeffrey A.Z. Zable*

The Fish and the Bird

One day a fish and a bird agreed to have sexual intercourse. The fish jumped out of its bowl and lay there on the table as the bird proceeded to mount her. Immediately the fish complained that the bird's claws were digging into her scales causing great discomfort. The bird immediately took offence at this confession and began flying around the room trying to work off some of his anger and frustration.

In the mean time, the fish was beginning to turn blue all over and was taking very deep breaths, one after the other. She tried to call to the bird for help but he was so busy flying around that he didn't even notice her. By the time the bird had calmed down, the fish was just lying there swollen with her eyes popped out.

Not realizing that the fish was dead, the bird flew down upon the fish and proceeded to have sex with her as he had done before. He presumed that he was doing an excellent job since he did not hear a single complaint from the fish.

When the bird was finished, he looked down into the fish's eyes to see what kind of response his lovemaking had made. The fish did not move nor make a sound, but just lay there with her eyes popped out. The bird interpreted the fish's reaction to be one of ecstasy. He imagined himself to be one of the most incredible lovers of the universe, for never before had he witnessed such a response to his lovemaking.—He began to soar around the room like a falcon full of joy.

The Feast   for Russell Edson

Two men are cooking up a beautiful woman in a huge pot.
The first man is telling the second man how he plans to
start with her arms when she is done. The second man then
tells the first man how he plans to start with her feet
when she is done. The first man tells the second man that
after he has finished with her arms he plans to go for her
mouth. The second man tells the first man that after he is
done with her feet he plans to go for her knees. The first
man tells the second man that after he is finished with her
arms and mouth he plans to have her breasts.—The second
man looks at the first man in surprise and exclaims, "You
are mistaken my friend! After I have done with her feet and
knees I am the one who will partake of her breasts!"—
The first man looks at the second man with a menacing expression on his face, and declares, "You are wrong my friend! I
am the one who will enjoy her breasts after I am finished
with her arms and mouth, and if you try to cross me I'm
going to smash you to pieces!"— The two men stare at each
other with violent expressions of anger, and then suddenly
strike each other at the exact time. They begin to beat
each other with fists and feet until finally both of them
are lying face down near the pot, bloody and unconscious.—
The woman, seeing that things are now perfectly safe, jumps
out of the pot and shakes the water from her body. Slowly she
begins to lift each of the men and drop them into the pot.—
She sits down in front of it wondering which part of their bodies
she will start with when they're ready.

Arthura Okamura

An Old Story

A woman was chopping down a man when all of a sudden she noticed a bird nesting in his hair.— She laid down her axe and gently lifted the bird from its nest.— She was very pleased to see that the bird was mothering three beautiful eggs.— She put the bird back where it belonged and once again started chopping down the man.— She was about three quarters of the way through him, when she heard some strange sounds coming from his head.— Once again she lifted up the bird, and was very surprised to see that the eggs had hatched.— Looking closer, she saw three little men lying there, blinking and rubbing their eyes and crying out for milk.— Realizing that a bird could no more feed men milk than a coke machine, she quickly took off her shirt and bra and began to raise her bosoms up to the men's heads.— Immediately the little men began to drink and birp, and within minutes were fast asleep on top of each other.— The last thing that the little men remembered hearing was, "TIMBER!"

SITUATION NO. 2

Daniel Roebuck

I was out walking and this guy stopped me and asked for a cigarette. He told me his name. He said you never heard it and I said I never heard what. I gave him a cigarette and he said this looks okay here all these stores, I wonder if they'd be easy to knock over. Well, I said, I really couldn't say. He said hah remember that big job in the jewelry store. That was my brother. I guess you'd have to be pretty good I said. Don't worry about that. You got nice eyes they match your shirt he told me. He pulled up his shirt quick to show me the gun stuck into his pants. He said you wouldn't want nothing to happen to them eyes would you. I smoked my cigarette. I don't know what you're talking about. He smiled and said he liked a guy who knew how to take care of himself. I figured we parted friends. I don't know if the stores were easy or not. I like a guy who knows how to take care of himself.

I was out walking and this guy stopped me and asked for a cigarette. I'd slid out of the lacquered sheets to go for a walk. The sky was sharp. I thought it was sharp enough to open a cut in my cheek like sleet.

My grandmother had said, if you always say no you'll be safe.

I took a deep breath; it made me feel like I was in the movies or on tv. I let the breath out slow like doldrous music over endless credits, my name was lights on the screen quivering; I sneered at my image.

She'd schooled me well in the overwrought gesture: I turned to simper at the moon. On the other hand, I often pretended to be asleep, trying to create drama out of absence, when she called me form another room or slid into bed.

He told me his name. He said you never heard it and I said I never heard what. I was a genius at forgetting. I'd forgotten when we smiled svelte words at each other and only remembered our stiletto nails. Or was it misrepresentation, facts all neatly sliced and incorrectly filed, so that they'd disappear. I saw her face disappearing against the pillow, as if the pillow were a salivating mouth, swallowing it.

She said my imagination was just false memory and the pulp of fear. Of course I didn't remember his name. Her name was stuck in my throat like a slow blunt arrow.

I wondered if how we were was an aberration, or part of a rigidly pendular scheme, in an arc like meat. Either way I'd forgotten all the connections.

I gave him a cigarette and he said this looks okay here all these stores, I wonder if they'd be easy to knock over. Well, I said, I really couldn't say. I looked at the stores all rouged and glittery. Even though they were closed they squealed like twelve year olds at a party, whose pubescent glamor was already beginning to decay.

She and I had had so many conversations that were just ways of telling each other we'd be hard to knock over. Not that it was true; it was a lot of bluster and fear. We threatened each other during tv commercials, which encouraged us. Things were cheap in the realm of stores and tv.

He said hah remember that big job in the jewelry store. That was my brother. I didn't know what he was talking about but I nodded and made a face like sure, I was considering pulling that job myself.

Cigarettes prolong conversation. Commercials tell you how to handle difficult situations. It was just the way I looked at her sometimes: I know *exactly* what you're thinking; when in fact I knew least of all.

I thought about my grandmother's theory of safety. Risk was a form of imagination, although I had a tendency to romanticize what was merely false.

Once I had told her that being with her was something like being in a department store. She nodded as if she knew what I meant, although I didn't.

I guess you'd have to be pretty good, I said. Don't worry about that. I wasn't worried. Or I was plenty worried, if not precisely about that. She could be sleeping, or gone, or dead. A sign flashed on and off in one of the stores. I tried willing it into portentousness. Of course I hadn't wanted to close the door. Walks were a faded ploy and the streets always slippery cold like inside an allusion.

I didn't dream that, I told myself. In the dreams I was always wrapped up and couldn't move or under the water where why didn't it feel like I was drowning.

Or she could be sitting up, tugging at her hair, smoking a cigarette. I tried to think of a recommended antidote from the commercials. The dumb stores were closed anyway.

You got nice eyes they match your shirt he told me. I squinted hard and all I got was a blur. My head buzzed test patterns.

He pulled up his shirt quick to show me the gun stuck into his pants. He said you wouldn't want nothing to happen to them eyes would you. A conspiracy of silence. I couldn't remember when I'd last seen a gun that wasn't on a screen, a page, or a cop, although I was convinced they were all over. She'd talked about a gun under her pillow once, in her strange voice as if it were only a lost tooth. I couldn't picture it. I told her it would be sure to blow apart one of our sleeping heads during the night. She raised her eyebrows and said she was mostly kidding.

I smoked my cigarette. I don't know what you're talking about. I thought his head bounced up and down like a ball, but it was perfectly still. It was my eyes, how the comparison to a functioning camera didn't work. My eyes were strapped to the raft of my head, floating, insulated. I said the opposite of what I meant, heading for safety.

I looked for her face where it was swallowed by the pillow and heard my grandmother calling. No one was on the streets. The stores all told lies. Whether or not a situation demanded action, inevitably I took none.

He smiled and said he liked a guy who knew how to take care of himself. My head was cracked, filling with dust. I held onto a blank smile. Very knowing and nonchalant.

All heavy time and times congealed inside me; I was crushed by the blank smile. She was no longer willing to make the effort to try to understand. I could hardly blame her. She'd been impressed at first, or something like that. Then she tried to pin me down: facts, you're generalizing, you don't make any sense, you can't make up your mind, you're weak.

I felt like I was put together with pins and thread.

I figured we parted friends. Suddenly the value of strange casual conversation was enormous. I put all my faith in it, my faith was the air in a paper bag, I slammed my hand down and it exploded.

I don't know if the stores were easy or not. I like a guy who knows how to take care of himself.

A Philosophy

   there are times we disappoint each other
                 no blame
pointless to blame an oak for its acorns
because I yearn for apples
and if I mistake an oak for an apple tree
better to clear my eyes
than to grow angry
willows make good baskets
and poor walking sticks
if we disappoint each other by being what we are
                no blame

a philosophy easier to apply in the abstract
                than when needed

—*Lenore Kandel*

Ed Buryn

PASSION'S DURATION
Thomas Farber

Y ou should have seen them when they first met. God, they couldn't get enough of each other, making love until all hours of the night, on the phone if separated for even several hours, thinking of each other whenever they passed beyond the reach of still another sweet embrace. This was, truly, the passion of which the poets sing.

Skeptics, rest easy: romantics, stand and be counted! Ponder this: sustained passion is similar to perpetual motion, a perfection inexorably denied by some almost negligible friction. Why talk of blame, character, choice? There are laws. We can even attempt the thermodynamic, something like, "Passion is inversely proportional to familiarity." (After an initial period of grace, of course.) Think of it. Perhaps passion, like other forms of energy, can be neither created nor destroyed. Merely found, and lost. Perhaps also this constant amount of passion in the atmosphere is not quite enough to go around, must therefore be kept moving for the benefit of the species as a whole. How else would people ever get together?

If such speculation seems too mechanistic, let us examine the facts. To begin with, our lovers so exhausted themselves in bed that one evening she suggested they sleep apart a night, just to get some rest. She put down several cushions on the living room floor, brought out another set of sheets, and was dreaming—of him—by the time her head hit the pillow she'd taken from the double bed. He, too, quickly fell asleep. But, waking several hours later, cold north wind shaking the windowpanes, he got up, shivering, and came into the living room. He told himself it was wrong to disturb her, yet, grinning to imagine himself Prince to her Sleeping Beauty, brushed her lips with his until she reached out for him. They woke, very late for work, utterly exhausted.

The next night they tried it again, but this time he promised to let her sleep. In the morning both felt better and made love with renewed vigor. Several days later, sensibly enough, they purchased a second mattress and springs. If they spent nights in the same bed, both agreed, their passion would utterly consume them.

How to calibrate change? This nocturnal separation was merely an attempt to be reasonable, an acknowledgment that there could be too much of a good thing. Led to nothing more serious than arguments about who would come to whose room to make love; who, passion spent, would have to traverse the ice-cold floor back to an unwarmed bed. No, though a bit odd, this arrangement was hardly remarkable, nothing to misconstrue. On the other hand, after they'd been together nearly a month he developed an allergic reaction to the hair of her blue-eyed Samoyed. His own eyes began to water if she embraced him without washing her hands after petting the dog. Sometimes, approaching him for a kiss, she'd realize that her hands weren't clean and would have to control her impulse. And, not surprisingly, sometimes she just didn't feel like washing her hands yet another time. He went to the doctor for shots, never blamed her or the dog, but there it was, a small impediment to the direct flow of feeling. A little something between them.

Though their passion continued unabated, a constant miracle they created together, there were several other minor problems. She complained that he failed to clean the dishes carefully when he washed them. After starting to protest once that no one else had ever mentioned it, he caught himself, smiled, and promised to be more thorough. Meanwhile, he found that he didn't much care for her closest friend, who seemed to him to be displeased that they were living together. He even got into the habit of going out "on errands" when the friend came by. Though he never mentioned it, he was irritated by feeling compelled to leave his own house. But, of course, he'd been the one who'd insisted that she give up her apartment.

None of these problems, really, affected the great desire they had for each other. In this period, however, having experienced a series of debilitating stomach ailments (which, she noted, began right after she moved in with him), she decided to go on a strict organic and meatless diet. For nearly a week he shared her new regimen, eating large salads, much fruit, tofu, and various kinds of nuts, but then announced one evening that he'd bought a steak. Sitting alone at the dining room table after he'd broiled it, he ate with pleasure and noisily drank the juice from his plate.

That night he also told her that organic produce was just too expensive. Soon, accordingly, they purchased their food at different stores to prepare separate dinners. Initially, one cooked and then waited for the other to do the same so that they could dine together. But sometimes one of them had an appointment or a class in the evening, and, kitchen so small, the process of making two dinners took time. Before long they often ate separately. Further, she'd found that the sight of raw meat now made her nauseous, that even the smell was more than she could bear. Trying to be considerate, she said nothing, but of course he noticed that she'd come into the kitchen when he was cooking to throw the windows wide open.

In the ensuing weeks she purchased a number of books on organic diets and pored over them, occasionally explaining to him, for instance, the mucus-inducing potential of the foods he consumed. He was more sarcastic than he intended one night when he told her that even if organic food was becoming her metaphysic, he preferred to be an agnostic, if need be even to have his soul end up in some char-broiled hell. Sensing immediately that he'd gone too far, he tried to retreat to the solid terrain of fact, and pointed out that she still drank countless cups of coffee each day and smoked cigarettes. If he thought that demonstrating an anomaly would stave off her anger, however, he was sadly mistaken.

Her smoking was in fact a real problem for them. He'd stopped five years before, after a tremendous ordeal of depression and what he remembered as near insanity. He'd actually told himself before meeting her that he'd never live with a woman who smoked. The sad truth being that he still yearned for a cigarette and didn't need the stress of constant temptation. He finally asked her not to smoke in the house but relented when the winter rains came. Instead, he posted a no-smoking sign on his bedroom door and kept it closed all day.

She told him she understood, even tried to cut down on her smoking, but something about the closed door exasperated her. Often he'd return from work and, an avid reader, would stay in his room with a book for hours, keeping out of the living room both because she smoked there and because her Samoyed liked to curl up on the sofa. Sometimes, looking down the hall toward his bedroom door, she couldn't help feeling that he was closing her out.

The problem of her cigarettes was finally solved, after a fashion, when he began to smoke again. Soon he was up to more than two packs a day, while she found that her own consumption quickly doubled. Feeling pains in his legs and chest, he couldn't stop himself from blaming her, while she, now suffering from a hacking cough, insisted that until he began again she'd always smoked in moderation. Every few days he'd try to quit, growing more irritable as the hours without a cigarette passed, until, seeing her light up, he'd grab the pack, glaring at her.

Could anything else, one wonders, come between them? Well, yes. She frequently suffered from insomnia, while he had an obsession about burglars. Often her restless pacing in the middle of the night would snap him out of a deep sleep, his hand groping in the dark for the can of Mace he kept by his bed. Worse, perhaps, he played the flute. Badly. An accomplished pianist, she not only quickly abandoned the idea of accompanying him, but grew to dread his practicing, the same mistakes repeated over and over again. Finally, her checks often bounced. If she thought nothing of it, he, coming from a poor family, had grown up forever in shame about the dunning of creditors. He knew it was none of his business, but just to see her leafing absentmindedly through a stack of unpaid bills set his teeth on edge.

What, then, of their passion? Strange as it may seem, even after bitter argument they could sometimes work their way back to each other by making love. By this time, of course, it was quite different than when they met. So much to ignore; so much to forget, such deep breaths to take simply to exhale the rage. Just to decide who would come to whose room required careful negotiation and diplomacy, someone had to concede. And even then her hands had to be washed clean of the Samoyed, he had to brush his teeth to eliminate the smell of cooked flesh. Nonetheless, on rare occasions, so much dangerous terrain finally spanned, their loving would be almost sweeter than before. Tinged, now, with the fear of loss; with self-reproach; and, in the small hours of the night, once in a very great while, with a piquant sadness for all that had come between them.

RELATIONSHIPS 63

Ed Buryn

RIDING LEATHERS
Deborah Frankel

I slide off the pillion and kick the pegs up, and I'm watching my girlfriend lock up the bike, when this voice a few feet away says, "Can I suck your cock?" I turn my head to the left and here's this good-looking blond, he's wearing cutoffs slung down around his hips, the kind that are cut off so short you can see the linings of the pockets hanging below the ragged edges, and damn near see the tip of his cock too, since he isn't wearing any underwear, and he's got a bunch of keys clipped to the belt loop on the right side, and running shoes, and that's all he's wearing. Skinny, tan and muscular, looks about twenty years old.

I don't say a word, but I walk over pretty close to him. You understand it's just a sunny afternoon in the park, but I'm wearing full riding leathers; engineer boots, black chaps over Levis, brown leather jacket laced very tight at the waist with sleeve lacing and padded shoulders, padded black gauntlets, and a black fullface racing helmet because I'm not twenty anymore and I want to keep my skull in one piece. And shades under the helmet. There is no way this guy can read my expression.

He is looking at me hopefully. Now here is a dude who likes to live dangerously. He sees a strange biker in leathers ride up with girlfriend and he wants to know can he suck my cock. The girlfriend, by the way, is cool; she's standing over by the bike waiting to see what I am going to do.

I'm not saying anything but I reach down his cutoffs and take hold of his scrotum. He likes this. He's a couple of inches taller than I am; I'm not really tall but I stand straight, people think I'm taller than I am. I brush my other gauntlet over his eyelids a couple of times till he gets the idea and closes his eyes. Then I pull off my helmet with my free hand and drop it on the grass. I cradle the back of his head in my hand and press it down towards mine; our lips meet and for some reason he doesn't want to give me his tongue but then he gives it to me. I'm just squeezing him a little with my other hand, just a little. As soon as he gives me his tongue it's like a deep well opening up in both of us, deep and deep, down to my hips at least. I can feel him through the leather, he's very excited, I stroke him a couple of times on the balls and just once on the shaft and he comes, he comes so quick I'm not sure he's done. Then I think, don't be stupid, he's a young man.

I bring up the dripping gauntlet to his lips and let him/make him taste his own semen. I put my hand down on him gently, again, and we stand so close together. He opens his eyes.

"You're a woman!" So I am. Even with shades on it isn't too hard to tell. The bones in my face, the way my hair is pinned up. A minute ago he thought he was in heaven, now he wants to get away quick, but it doesn't take him too long to realize that a leather-jacketed stranger of uncertain intentions has still got hold of his balls. I spare a second to glance over and see how Carol is taking all of this. I don't begrudge her Wednesday night man and her Saturday night pickups, not to mention the knight who helped her put the motorcycle together out of old Triumph parts, so she would be in a bad position to get possessive right now. In fact, she's entertained. As I expected.

To return to my betrayed faggot. My one hand in his shorts, the other in my jacket pocket. "Let's say for the sake of argument that this hard object in my pocket is a loaded automatic pointed at your belly. And let's play this scene over again, this time with you knowing that this stranger isn't the stranger you were hoping for."

"Is there a gun in your pocket?"

"Maybe it's a bunch of keys. Want to find out? For the sake of the scene, let's say it's a gun. By the way, my girlfriend is good at watching everything and keeping her mouth shut, and she's seen me do weirder things than this, and she's a very convincing liar."

Maybe he doesn't think I have a piece in my pocket, but what's he going to do? I still have my hand on his balls. Besides, the leather is confusing him. He's used to taking orders from it.

"Close your eyes." I stroke him but he's limp; of course it's going to take longer this time. I slide my hand back between his cheeks and begin working the thumb in; the padded glove makes it thicker and slicker which is nice for him, and keeps my hand clean which is nice for me. Now that he isn't thinking about guns anymore, I slide my other hand up his bare chest. I love playing with nipples, and the welts and padding on the glove give us a whole new range of sensations. I don't have quite the span to get both nipples at once so I brush the other one with the lacing on my jacket sleeve. I work his tits lightly, then roughly, though it's hard to get a good grip in the slick leather. Now that there's no need to conceal my sex, I'm pressing my pelvis hard against his, and I can feel him start to swell up. I have my thighs around his legs as much as I can. Skinny-legged boy; I could almost make a pair of his thighs out of one of mine. I bring my other hand down on his genitals while my thumb is still hooking for his prostate and the fingers brushing his soft, smooth ass. I've got him tilted back with my legs around him and hands cupping and stroking him and we rock together. He starts to groan; it's like I'm wrenching an orgasm out of the flayed layers of his skin. I wipe my gloves on the grass. He doesn't want to look at me. He can't see my eyes through the shades but he can feel the intensity of my gaze. I think his idea of himself has been destroyed. Why? A man's hand and mouth are constructed exactly like a woman's. And leather is leather.

"You live near here don't you?"

"Yes . . . no . . ."

"I'm tired of standing up, aren't you? Let's go to your place and play a couple more games."

"I can't . . . "

"You got a roommate? I'll put the helmet on, don't introduce me, he won't know the difference. Forget about history. Now is now. C'mon."

"I wouldn't be able . . . "

"You didn't like it? C'mon. I can teach you somemore, maybe you can teach me something, huh? Use it later with somebody else, huh? Hey Carol, if I'm not back by five, leave without me." Carol walks over to us, smiling. She puts her arms around me and we kiss each other a long time. "Have a good time," she says.

"Enjoy yourself. C'mon, let's go."

66 CITY/COUNTRY MINERS

Ed Buryn

Bedtime Talk Between Two Six-Year-Olds

–What is dynamite, Papa?
–It is contained, and then it explodes.
–Like a bomb?

–Yes, like a bomb.
–But how is it made.?
–Dynamite isn't made. It's captured.
–Captured! Like an animal?
–Like a snow tiger caught in a net, its claws sharp
 as icicles, its eyes sparkling like cut glass!
 Yes, it might be an animal that lives its life
 in one brilliant instant; but do you know
 what dynamite really is?
–What?
–Lightning.
–Lightning! But how is it captured?
–In certain places, usually on mountain tops, there are
 glass castles that capture Lightning.
–Have you seen the glass castles, Papa?
–No, the castles are invisible.
–But, why doesn't Lightning shatter the castles?
–Because Lightning is so fast and brilliant, it sees
 right through the glass and doesn't know
 it has been captured. If Lightning saw the glass,
 if the castle thought that much of itself,
 then Lightning would surely shatter it.
–What happens after Lightning is captured?
–Well, I said that Lightning is captured, but it isn't
 really captured. The castle thinks it is, and that
 seems to be good enough. So, what the castle
 does is this: it funnels its thinking into wires and
 tubes and lightbulbs and dynamite and such.
 It gives Lightning many names and many uses.
–Does the castle make lightning bugs?
–Yes, you got it! And rainbows and waterfalls
 and colors, too.
–But, what is Lightning, Papa?
–The Sun. Lightning splits off from the Sun, and do you
 remember what the Sun is, what the old poet
 told you?
–The Sun and Moon are the eyes of God.
–That's right. And Lightning is His reins on the world.

—Frank Polite

Dear Christine

I enjoyed my stay at the farm.
I felt, completely at home there . . . which means
I didn't change a sheet or sweep a floor.
I did, however,

hang myself out to dry once or twice.

I helped myself to the house Chablis
and J.W. Dant. I opened a few cans of something or other
and popped a lot of health pills
that didn't seem to do much.

I also popped several Percodans (without N's
permission, of course) which did
seem . . . to . . . do . . . much.

I loved your raspberry jam.

My son was out a couple times. We gathered chestnuts.
I built a fire. A lady friend of mine showed up
now and then to lift the burden.
But mostly, alone, I read,

I wrote and rewrote an old story
and continued my ongoing wrestle with Eternal Lassitude.
"Lassie" has learned a few new holds over the years
that got me down, but I did throw off

my old foe long enough to finish "The Transformer."

Folly is a wonderful dog. But what does Folly *want*?
When she wasn't nuzzling the cat
that lives in the woodbox

she followed me around from room to room, up and down.
Or there she was, looking into every window
I was looking out.

Finally, it did occur to me: Folly wants Love.
Also, she wants to be in and she wants
to be out, *at the same time*.

Still, we got along, although
Folly's psychology is too uncomfortably close to my own.

All in all, it was a good month. Not hot. Not cold.
No snow, I stood around down at the pond a lot.
I walked into rows of winter corn.

Where did the fish go? I looked in, and in,
but I didn't see any.

Your mailbox is a Jupiter cup. Every day it replenished
itself with the stuff of this world, from places
and organizations I never heard of.

And causes . . .

I answered one in particular, an undeveloped country.
I sent them a dollar.

I loved Folly, the woodbox cat, and the fire . . .

—Your Friend Frank (Polite)

Blood

on 58th St.
off
Slauson

in front
of a
liquor
store

this black
kid calls
this black
kid

blood

as I go in
for the six-
pack

with all
eyes on
whitey

I'm bloodless

& I'm still
alive

—*F.A. Nettelbeck*

Trina Robbins

You're Getting Old, Andy Brumer

These fucking grey hairs
on my head aren't even grey
they're white!
Not the distinguished
businessman in suit
walking with sexy leather
attache case down
the street in a hurry to meet
another executive to talk about
more money,

but slippery soft
Robert Frost hair
contemplating delicate
almost Chinese snow fall
on windless Vermont nights;
the hair of a compassionate
doctor who has been humbled,
not hardened, by all he's seen.

This is not the boyish mane
of a professional baseball
manager who still dives head first
into second every time one of his
players does.

This is Weasel hair.
Weaver's hair.
Not white of flower hair,
but porcelain hair.
You're getting old,
Andy Brumer,
hair.

—*Andy Brumer*

RELATIONSHIPS 73

Joel Beck

From **Take Wing**

"i feel as if my life is about to take wing."
—david soloway

PSYCHE & CUPID

his wings are tinged with
sunlight!

> *dont be silly. he
> has no wings. he is
> only a man!*

his wings are tinged
with sunlight!

> *not long ago,
> i was held, just that
> way; so airily*

he flew away

> *nonsense! men
> don't fly!*

the gentle ripples of
her skin / from hip to
tummy / her arms en-
circling his wistful
face / his arm across
her breasts, so
lovingly.

the wine dropped,
forgotten. the lover's
arms so full.

her throat pale, &
sleek. his hand cup-
ping her cheek as she
lifts to kiss
his lips; the lips of
the one with wings, in

    the light.

here come the amer-
ican boys. green-eyed,
intense. they chew
gum thoughtfully as they
approach the luminous
statue of psyche with her
love. her winged lover;
her gentle lover, with
the airy embrace.

his arrows rest
in their quiver;
her heart pierced by
the invisibility of
love.

his wings erect,
his body lithe. who
would not reach
for him; who could
refuse the glory of
that moment, time-
less as stone.

continued

the only garment left is
loosely draped; soon,
soon, as he pulls her
to him, it will fall, &

he will be hers.

effortless as the
seine; inevitable
as sunlight

    (o, do
    not be deceived by
    his wings! he is a
    man, & has never
    flown! & with you, he
    has only the illu-
    sion of
             (flight!)

woman with white
skin, your touch cool
as marble, the
weightlessness of

    Wings!

(have you a husband,
psyche, whose heart
would break? have
you children, whose
home could collapse
for the sake of your
hot moment? or
are you the inde-
pendent type; the
kind who never answers
to nobody; who pro-
claims the rite of
the body to choose
its own mate

she looked right at
her lover, & she kissed
him anyway! that's
how beautiful he was!

& his hair is curly,
too. as if the rest
weren't bad enough, his
goddamn *hair* is
curly!

he has just flown to her
side! that's it! she
was sitting alone, & drinking &
all, & he whizzed down
to take her in his arms
as if she had no other
business in the world
than to let her breast
fall into his palm!

she didn't fall for
samson, you notice. or
caesar, or louis the
IV! o, no. some wisp
of a youth; some winged
creature! some man without
a proper robe, or
vestments.

some man who came
to her naked,
    as innocent as
she.

—*Alta*

Kristin Wetterhahn

Be Limber

You are the young
wild catastrophe
a bunch of yellow flowers
in a fired field

The morning dew
gives weight to the ashes
& a freshness
to your closed bloom

In this burnt patch
of moonless patience
you dream
the color of caution

wave uneasily
in the warm
rising valley air . . .
Are you telling me something

are you swelling
minutely
for dawn
Is the moment of East

your infant memory
your dumb signal
in the logic of roots
Will you remember

how I saw you here
Will you stay tenuous
on this passover slope
A yellow banner of risk

humble for the light

—John Mueller

Barstool Cowboy

Barstool Cowboy
how did you become so wise
did you learn to judge the world
from watching shifty eyes?

Barstool cowboy
that's not a halo on your head you know
it's just a dime store stetson hat
Do you really think it can
make you look like a man?

Cowboy, you sure do seem proud
talking so loud
for the sake of the crowd
bout rustlin the round up
bartender turn the sound up
on the radio
cuz we don't wanna know
Cuz cowboy your last shot
was the one that you just bought
Barstool Cowboy.

Barstool Cowboy
those boots are wearing thin
the places you have been
have led you nowhere
barstool cowboy
do you think you'll ever change?
you've been riding this range
too long.
And the horse that you ride
is just cheap naugahyde
barstool cowboy.

—*lyrics by Barbara Suszyeanne*

Waterfall

No one is there and you are not alone.
You are breathing with an enormous face
still lit with the light of dead stars.
Your skin is the color of autumn

ripe, fallen, red moisture surrounding your lips.
Kissing the long lost twist of love
you are able to make water laugh
a rainbow when your body hits the air.

—*John Mueller*

WESTERN HAIKU

VOLCANO

I could live
    In quiet again
But I won't

Remembering Nixon's face, I kick a pebble

Under the shadow of the reactor
    I clean my glasses—
Morning-Sun!

DURANT & TELEGRAPH

On opposite corners
    2 guitars
  out of tune

I wash my face
again and again
but I still don't see

Sitting by the pond,
    there are too many splashes
To separate

In the white fields
    of my allegory
A Morning-Glory springs up!

MAGANY

Watermelon kiss:
    for lunch—
        you

—*William Garrett*

84 CITY/COUNTRY MINERS

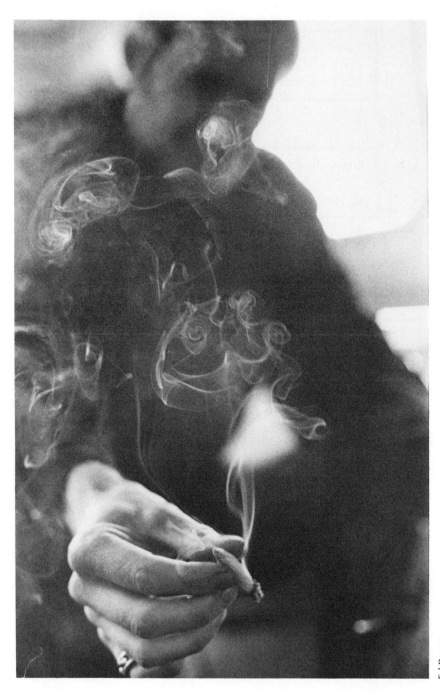

Ed Buryn

Dope Poem

one of the things about dope is
you can never take a piss because
the bathroom is always full of people
shooting up
and you can't stir your coffee while you
wait
because all the spoons are gone
and when you find them hidden
behind the bathtub
you have to wash the soot off and
they don't taste right anyway
and if you take your turn and
sit on the john and
roll up your sleeve
it's only a hot flash
in a cold world

—*Lenore Kandel*

Dealing with Stress

Sunday! Sunday!
There is another
world, I know
there is, because
this is it, these
shut down storefronts,
the morning sun which
butters me against
the window, things
priced, but not now
for sale, set deep
in trembling glass. I am
being taken
to Sunday school
and I don't even mind,
because in this place,
at this pace, the senses
close down and slowness
drips over me and sticks,
and is all
that will remain
of this day.

Shaking Off the Small, Compounded Madness

Pinholes of darkness appear in your eyes
letting the visions escape you with a hiss.
Within a continual instant of awakening
everything you see is flying away
at incalculable speed, at no speed at all.
Here is pure pleasure,
here are the huge tanks growing empty
and the world snapping back with a barely discernible rustle
to where it has been all the time.

In Wolves Clothing

The eye grows hard with invention of edges,
the face takes on knowledge of what must be:

acceptance of the living room,
the door beyond which there is no breathing.

In youth, when everything had to be proven,
the best possibilities were all discarded,
insufficient evidence.
                      Here, at the endpoint
of reduction, tongues cluck, the news
passes for scripture. A casual glance confirms,

experience is the best teacher.

—Bruce Hawkins

Sundown

Now where we stand in purple twilight
with the ochres of tones and the cinnebar
alchemy of the land given under eyelids—
russet the dreams of flesh, tawn roan
as soft shale or the loam of the inlet
fingers; bodies with that music in them
and of the masses of the earth-given
and sun-driven sweat come
to iridescent life across the wrist-wipe
of a brow of smile.

It is the field of
the berry and vine, it is the green
heads of the army of lettuce
scrutinized by the sunset, it is
the gender of the exasperatedly simple
and the plumtree and the cornstalks.
The reddening spun coronas play upon
the levers and the gears. Life shambles
with its dignity of history, positions
of posture and cordage: *We take root*
is almost an imperative to be
repeated til it echoes through
the body caverns. We take root the way
turnips or heroes, when they are
humblest and therefore typhoons,
storm in themselves so that
horizon be clear as
a line drawn by water.

This is the land the music of a man
makes into face the other side of
the face of artifice. Artifice continuing
but with its style of content, its content
of burden and ripeness, the animal
song of the summer roads of profile
and rut mated with its youthful
cataract beside a tree overlooking
the portrait forever. They are
given me. Us is.

We are given food for the night song
of a thousand crickets of sweet
melody. I find my I only as it is
born in you who take me through
these dark journeys pouring stars
out of the sky you've found,
which I praise and wear
like a cordilliera
in the shape of a heart
as, naked, we ride
each poem to its necessary river.

—*Jack Hirschman*

Love is an Art

love is an art for angels
and we are human, you and I
fallible we are, and fragile
and therefore more than perfect
we take such risks who leap across the void!
perfection is static paradise
but we are human, you and I, and so we dream
and cast our dreams before us
extending our fingertips beyond the finite edge
to brush that certainty of ringing bliss
that resonates our dreams
impelling us to be that art
which angels strive to emulate

—*Lenore Kandel*

PLACES

*If there are no whales,
what will you show your children?
Bones in museums?*
　—Garry Gay

NORTHERN CALIFORNIA
Raymond Dasmann, talking

I think of Northern California ecologically, in terms of the whole complex of mountain ranges, valleys, vegetation, animal life, that spreads through it all and ties it together. The chaparral, for example, is a unique kind of vegetation you don't find outside of this state. It's primarily found in the Sierra foothills and along the coast ranges. It's characteristically Californian. You find it in Southern California also. It's not really something that divides but rather ties Northern and Southern California together. The Redwood forest *is* uniquely Northern California. It starts in the middle of the Santa Lucia Mountains and runs up into the tip of Southern Oregon. The complex of pine and other conifers in the Sierra Nevada, which is shared with other states, also adds to this feeling of unity in the California bioregion. I think, too, of the animals; the deer, the birds, the coyotes and the grizzly bear that was once here and isn't here any longer. I miss these when I'm away from the area. They belong to this region and are part of my way of relating to the world.

The historical background here is important to me. The fact that this land was occupied for thousands of years by the Native Americans and they've left their imprint on the land. You feel somehow that they're still here. You'd like to know more about how they related to the landscape, what were their sacred places, what were the important centers of activity. When I look out around the Bay Area at Mt. Diablo and Tamalpais, Mt. St. Helena farther north, and the high tops of the Santa Cruz Mountains, I think these high places—you could look at one from the other—must have been important to Native Americans. But how important? What were they? I don't know. I would like to know more about that.

The imprint of the Spanish is another thing that makes California unique. Everyone grows up in this state kind of halfway speaking Spanish just because there are so many place names and common words that we use that are directly from Spanish. Also, the architecture, the way the buildings are designed. So, you start with an ecological framework, add this historical and prehistorical background to it, and get this feeling of a unique place that you can't find anywhere else. You certainly don't find the same feeling when you go over to Nevada, or when you go up into Oregon. When you go to Oregon you begin to get the north-coast-logger feeling, much more strongly than you do anywhere down here.

How different Northern California is now than at the beginning of human habitation in this place is an interesting question. The Indians really didn't use all that much of the state. Their centers were more localized. They may have moved out in hunting trips or traveled to visit other peoples. But for them a lot of the state was just a background for activities rather than a place that they used intensively or occupied with villages or structures. The Spanish were also very localized. When you think of the areas that they actually settled in Northern California, there weren't very many. I often wonder what would have happened if the Russians had stayed longer. Because they were bringing a different influence from the north down into the state—a different style of building, one in many ways more suited to the northern parts of California. But until U.S. settlement much of the state was just very wild and not permanently occupied. The original U.S. settlers spread widely over the state. They tried to live almost everywhere. You see old pioneer cabins and remains of pioneer activity like goldmining debris, old dams, flumes and waterways everywhere. In places where people don't live at all now. It's as though people spread out into every canyon and tried to make a living, then retreated back to more secure areas as time went by.

The impact of recent civilization has been, of course, enormous—in the sense of filling up certain areas. The San Francisco Bay Area, for example, with its housing and human habitation. And yet, when you fly over the state now, you see how localized settlement still is. There are still huge areas that are hardly occupied. They get a lot of pressure in the summertime from people trying to find outdoor space but they're not permanently occupied, and there isn't any heavy human imprint on them even today. I think that this is one of the things that people outside of California don't seem to realize about this state. They think of it all in terms of the San Francisco Bay Area or L.A. They don't think in terms of wild country. In fact, they don't even want to believe it when you tell them that most of the state is still relatively wild. Still, the demand for resources, the reaching out for everything from minerals to timber to grazing land, exerts a lot of pressure on even the wildest areas. One doesn't feel that there's too much space. As far as real wilderness goes, where you can really get back away from people, that is not common at all.

The Central Valley

If you look at the maps of original California and original California water and vegetation you see that the Central Valley was a valley of very extensive fresh water lakes drying up into marshes in the dry season. I'd say that channelization of water in the Central Valley has been one of the biggest changes. Those areas that were marshes or lakes are now farm lands. They're irrigated farmlands and the water's channeled to particular uses. Drying up the Valley has been one of the things we've done, in the sense of restricting the area in which water could move freely. Providing irrigation channels, that direct water to specific targets, has had enormous impact on the Valley as a whole. Very few areas haven't been affected. The vegetation is changed almost completely. The original California prairie is mostly destroyed and replaced by either agricultural crops or by the introduced European annual grasses that have taken the place of the old California perennials. The animal life has almost vanished. The Central Valley was the center of huge herds of elk, antelope and deer. Now, both the elk and antelope have gone, except for a few little places where there are some elk left. Deer have been greatly reduced, grizzly bear are gone, beaver have been cut back. Almost any species you name, except the agricultural pests, have decreased. In the past 40 years a pattern of subsidized agriculture has been imposed on the Valley. It has increasingly been maintained by the general public for the benefit of a relatively few big corporations in many parts of the Valley.

A lot of money that's made in the state from agribusiness goes outside the state. It doesn't stay here. The support of people has dwindled. There used to be lots of people working on the farms or owning small farms and having an agricultural way of life that was quite pleasant. Now, you largely have big land holdings employing seasonal migratory workers at minimum wages and not providing a suitable habitat for them or a suitable way of life. It's true that agribusiness is a 14 billion dollar industry—if you measure it in gross terms—but not much of the gross I think goes to the individual people in California.

I wouldn't want to bet on how long this subsidized agribusiness will be sustainable. It seems to lead to depletion of ground water and resources, to ever increasing demands for more surface water to be brought in from elsewhere. It's based on water-wasting crops to a very heavy degree, on energy-intensive agricultural methods that can't be sustained indefinitely, on heavy use of poisons of various kinds that are impacting the soil flora and fauna as well as the larger animals and humans. I see very little about it to feel happy about. Agriculture could be made sustainable. The Central Valley could be an agricultural paradise for a thousand years, but not under the present methods of use. I think we have to get people back on the land who actually care for the land and who will use farming methods that can be sustained. Ones that don't demand enormous subsidies of water or energy and don't rely on toxic materials as a means of battling against insects or diseases.

The medfly program has, for example, been mishandled. It is now being mishandled more severely than ever, in the sense that I don't have any belief that this aerial spraying program will actually eliminate the medfly. Aerial spraying I can see working within the limitations that any sort of poisoning works—on relatively uniform croplands and field areas. But when you start spraying urban environments with all the shields and buffers and houses and eddies set up by all the man-made structures, the chances of getting uniform application of malathion on the ground are very limited. The chances of missing the target areas are, I think, great. Even if they succeed in getting malathion all over the ground where it is needed, the concentration of it would be less than could be achieved by hand-spraying, spot spraying. I think they should have started in earlier, made people more aware of the problem, worked much more closely with the local people to get the fly attacked right where it lives. To me aerial spraying is just an indication of inability to cope with the problem. You can't get rid of the termites, so you burn down the house, that kind of thing. The cumulative effect of aerial spraying could well be the development of malathion resistant medflies, which will then go on to do greater havoc.

Still, the biggest agricultural problem is finding ways to get people back on the land. There are lots of people who would like to farm. But you can't go out and buy farmland now unless you're very rich. You *can* buy marginal land. A lot of the important farming for the future is probably taking place on the margins of the main agricultural areas. The small-scale farms and experimental organic farms that you see cropping up all over the foothills around the edges of the Valley, and around the edges of the big agricultural centers, are probably the best hope for the long-term future of agriculture. There's also a big need to develop farm co-ops that bring a lot of people together and enable them to get pieces of land, really first-rate agricultural land, that they can then begin to take care of.

Cities

Of course, most people will probably continue to live in urban centers. The thing is to make the urban centers more livable, to make them more self-reliant than they are at the present time. We have to begin to connect them once again to the lands that support them, to the farms, forests and fisheries and such. We have to begin to restore the flow of energy and materials from land to city and back to land. Right now, the cities tend to be the black holes in the system. They drain energy and resources off the rest of the land and dump them into entropy. Getting human wastes separated from industrial wastes and back to the land is a major thing that must happen. Otherwise you have a constant drain of nutrients, which means that the agricultural lands in the long run will not be sustainable.

In a more direct human habitation sense, the building up of self-reliant neighborhoods is a major concern. However, I think you have to be careful about self-reliant urban houses as such—each one with its own windmill, solar collector, and little Clivus toilet, living as a unit all its own apart from its neighborhood. I don't think that's the way to go, in the long run, because it probably involves a waste of materials and energy. Local, neighborhood-based or small town-based systems for energy generation and sewage reclamation are the way to go. Local sewage collecting systems could, for example, generate methane which, in turn, could be piped into the gas lines that go around each neighborhood. Processed organic waste could be used for neighborhood gardens and the surplus exported back to the agricultural lands outside the city. This could also work on other energy systems. We could have more general solar heating of water and more generation of electrical power by neighborhood wind generators and solar cells. Disconnecting from massive area-wide systems will be difficult. But in the long run that's the way we'll have to go.

Bioregions

Northern California is a federation of bioregions. There are distinct areas, each with its own particular flavor within the general Northern California region. There is, for example, quite a difference in attitude, and in the ways people relate to country, between say Bakersfield in the southern San Joaquin Valley and Redding under Mt. Shasta in the north. This relates to differences in the rainfall, the vegetation, and the history of each place. Redding traces its origin back to the Gold Rush days. It also relates very much to logging in the coast ranges and Shasta region. Whereas Bakersfield ties much more into the old Spanish land grant days, to huge cattle ranches and then to oil which became dominant once cattle ranching began to fade out. Oil and agribusiness now go together in Kern County. Agribusiness is being carried out by oil companies. So there's a different atmosphere in the two ends of the valley.

If you start really looking at things more closely, you see even more distinct units. My home area, Santa Cruz, for example, is cut off by mountains from San Jose and the Bay Area. It's also separate from the Monterey-Carmel area. You could almost use the Salinas River as a boundary between the two. Santa Cruz has a very definite quality of its own. The heavy influence of Redwoods for example—the Redwood forest is a dominant feature here. The marine terraces with their ever increasing elevation as you go up on the coastal mountains. The potentially high agricultural productivity. The ocean is a very dominant feature in everyone's life, everyone relates to the ocean in some way or another. We have our own watersheds, the San Lorenzo Watershed, along with some of the smaller coastal streams. Santa Cruz is potentially an area that could be independent of other water sources, it could be independent of other energy sources, it could be an even more distinct unit. It is a unit that people relate to in a way that they don't relate to a broader Northern California. Santa Cruz is a place where people live. I don't just mean the city, I mean the county, the little Santa Cruz bioregion. Whereas Northern California is too big to identify with. It contains places where people live but other parts that they only visit.

When I go from Santa Cruz up to Nevada county (where we have land on San Juan Ridge) I think about the fact that county boundaries don't usually mean anything. Up there the watershed boundaries *are* important. I think of the Yuba River people as people living in that particular environment along the west slope of the Sierras, from the valley up to the top of the range. That's a unit that people can identify with. I see that as quite distinct from, let's say, the Feather River complex, which is just the next major watershed north. Because the Feather River drains from a volcanic area and is very influenced by lava flows and other vulcanism. Whereas the Yuba River is old Sierra granite without too much volcanic activity affecting it.

The Yuba watershed is an area that people relate to directly. They live there, they know it, they know all the activities related to it, the problems associated with it. Whereas it's very hard for a person living in Nevada County to relate to the problems of Humboldt County or even Santa Cruz. They're very different kinds of problems. Nevada County itself is too artificial. It goes right up over the top of the Sierras and takes in the east side around Truckee. There you have an entirely different situation. The barrier of the Sierra crest really draws a line, cuts off the east side from the west side. East side people don't want to relate to Nevada City on the west side. They don't really feel that they're part of Nevada County. When winter comes that Sierra crest is a pretty formidable barrier. People who want to go into town, who live on the east side, probably go down to Reno. They don't come over the top and go to Sacramento to reach a city. So, the vegetation is different, the ways of life are different, the relationships are different on both sides of the Sierra.

However, one of the things that I get a little concerned with is attempting to define bioregions entirely on the basis of watersheds. Sometimes the watershed is the important unit, but it isn't always. Santa Cruz and Monterey are really quite different regions, different communities both in human terms and in biological terms. But, I picked the Salinas River as a logical boundary *separating* the two. I might pick the Pajaro River as more logical in some respect. But you can't say the Salinas River watershed makes a good bioregion because the Salinas River watershed takes in far too many different kinds of country. The people, who live on the west side of the Salinas River watershed in the San Lucia Mountains, have a different kind of life from people on the east side in the San Benito area in the interior coast ranges. There are very distinct differences between the San Lucia unit and the interior coast range units. And these aren't watershed differences. The Salinas River may drain them all, but the San Lucias are a distinct block of mountains with their own unity. So, I think you have to be flexible in deciding what are your bioregional boundaries and probably the important thing is how do people relate to that land, what do they think of as their space, their home area.

Sierra Foothill Footnote

Sitting on a stump
beside the broken trail—
Juniper, wild bracken,
and forest music all around;
the strong rising trees
lazily tossing galaxies of shadows
all along the ridge,

       I broke out laughing!

Any animal who's hiked alone
is familiar with this ecstasy.

—*Don Keefer*

Rock Outcrop on Beatty Ridge

If you sit too close
you cannot see it.
Step back
say 10,000 years
when its gaping mouth
was just a smiling crack
or farther yet
when it rose up
in a slow shock.

With another step
it is a clean slate
of beach sand,
a different one
each day.

Step forward,
it howls silently,
hidden from the sun.
An empty house.
A singing skull.

—*Jeffrey B. Wilson*

Joseph Carey

LETTER FROM MENDOCINO
Susan Pepperwood

It has been nine years since I came to live in Mendocino County via the Berkeley of the late sixties. Raised in New York City, without even a backyard tree, the back-to-the-land movement was truly a visionary one for me. What I saw for myself in my first acid trip, fortuitously taken on the Mendocino coast, was perhaps, a vacation home in the woods—but certainly not the full fledged poverty stricken hill-hippie lifestyle I chose after a summer of struggling to build that 'vacation home.'

Rather than returning to my Berkeley apartment and job in the fall of 1973, I held a huge garage sale, gave away my velvet and lace flea market dresses and my library of graduate school texts. I packed my jeans, thinking I would never wear a dress again, and moved into the unfinished, plastic walled vacation home—a cabin barely 15x20, lock stock and barrel.

The reason, I told my amazed city friends, was "it is just me and the universe up there." No middlemen on which to blame failures, no way of rationalizing defeats, no politics to compromise dreams. The purity of living in direct contact with the elements in a society free of governmental interference had overwhelmed me.

I was not alone. In 1973 the back-to-the-land movement was at its height. I had moved onto a 5400 acre converted cattle ranch in the foothills outside Ukiah which had been subdivided by a Berkeley visionary-turned-real estate agent. The land had been sold almost entirely to 'alternate lifestylers' tired of city living, and none of my two dozen neighbors were more than ten years either side of my own age of 26. Those first few years we worked together, learning and sharing carpentry skills and organic gardening techniques, living on our savings, hoping that our visions of self sufficiency on 40 acres would become a reality before the savings ran out. We tore down old barns for lumber with which to build our houses, we baked our bread, tended huge vegetable gardens, canned our winter food, and were as poor as we were happy. Isolated from and ignorant of the residents of the country we had moved into, we had no way of relating to the 'rednecks' and 'straights' who seemed to make up the rest of Ukiah. Our town trips were limited to the health food store, the auto parts store, and the laundromat. We had left politics when we left the city, we were responsible anarchists, and we related to the courthouse only to pay our taxes and our parking tickets. Our neighbors were our only friends.

At first we loved our isolation. Many of us were young couples, and thought we would be in love in the country for the rest of our lives. "I'll light the fire, you put the flowers in the vase that you bought today," had sung Crosby Stills and Nash. We cut firewood, grew enough flowers to fill our homes, and found clay to throw vases in seams of the hills in which we lived.

But isolation did not last long. It took just over a year for Mendocino County's conservative political structure to become aware of the huge influx of people moving into the country's hills. They knew nothing about us: few had jobs in town, many wore strange clothes and had (too) long hair, and almost none had purchased building permits before building their houses. In the winter of 1974, the Mendocino County Grand Jury ordered county officials to "seek out the violators" of building, health and safety codes, and a task force received funding to begin red-tagging (condemning) the non-permit homes we had built.

But although those "violators", survivors of the radical sixties, loved this county and the peace it induced, we were not about to see our houses destroyed: we had built them safely, and on our own property. We felt like victims accused of a victimless crime, we felt wronged, and we knew how to organize. We wanted to live free of governmental interference, but the government was trying to tell us "no." Within weeks, the first red-tag casualties had formed *United Stand*—an organization devoted to legalizing safe housing, irrespective of the Uniform Building Code.

It has always seemed significant to me that a group which now considers as its own two out of five county supervisors as well as the district attorney was born at a meeting called to protect itself from extinction by that same county government. Perhaps one of the reasons we here in Mendocino County are so strong is because the issues we have chosen to fight for are not just political, but have to do with survival. Our sense of community grew from our basic right to live as we choose in our own homes on our own land.

Although we organized, we did not want to get involved in confrontation politics. The county officials who were the task force set against us were invited to that first United Stand meeting. They had no solution to offer us, but they advised us to speak to the board of supervisors about our problem. The possibility of using the political process itself to achieve our goals was a revelation. The original building task force actually was a blessing in disguise—it forced us out of isolation and into a place in the community.

United Stand put us in touch with each other. Isolated homesteaders found at that first meeting that there were hundreds of folks living just as they were, with the same needs, the same dreams and visions. We found ways of working together, we formed a community center, opened a food coop, established our own library. United Stand began a newsletter, and when that newsletter was incorporated into the Mendocino Grapevine, a newly formed alternative newsweekly, the newspaper's circulation more than doubled.

Alternative businesses grew up. As people became better at their homesteading, they found that the skills they acquired could become a means of earning a living. City salesmen turned into country carpenters, office workers became community organizers, nurses became midwives, waitresses opened restaurants, artists and craftspersons found markets to sell their crafts. *Made in Mendocino*, a county arts cooperative was formed. Theatre groups, printing presses, a film theatre, an alternative health care center was organized. Alternative schools opened. Because people were doing what they were good at, because they were working with and sharing their skills with friends and neighbors, they were often successful. Money stayed local, and many skills were traded.

The network begun by United Stand has been growing for seven years now, and links every corner of the county. The Ukiah Community Center runs a 24 hour crisis line and an answering service for dozens of businesses. A huge file in the center lists hundreds of people and the skills and services they can perform: if you need something, you can check the file, call that person and perhaps work out a trade. Trading has become a standard practice here, where people have learned to trust each other: a doctor treats a lawyer with a sprained wrist, and is paid later, with advice on a legal contract; a veterinarian heals a sick cow in exchange for the sour cream the cow produces when she is well again. A baby is born with a midwife's aid, and; the father—a carpenter—helps the midwife add a room onto her house.

Networking extends into the hills. Much of Mendocino County is still without telephones, but communication is carried on in remote areas via CB radio.

CBs were first the domain of the truckers, then the four-wheel drive crowd caught on, but now the majority of CB systems are sold to rural homesteaders. At first CBs were purchased for emergencies, particularly fire warning systems, but they were discovered to be useful tools for information exchange as well. Residents in a certain area will decide on a station and turn on their sets at set hours daily. We arrange rides to town, make plans for work parties and dinner parties, find baby sitters, exchange herbal remedies, and spread the news. Some even fall in love. An interesting social aspect of CB communication is that it is public, word travels fast, much faster than it could by telephone. Communities are united more easily.

Now, seven years after that first meeting which united us on more levels than saving our homes, we are the landed left. Our homes are nearly completed, rooms added to cabins to house the children we have borne within them. Established water systems feed fruit-bearing trees we planted in the last decade, wisteria and roses climb past our windows. Solar energy systems provide light and hot water. Our survival systems are fairly automatic, and we have energy for larger issues.

The power and concomitant responsibility associated with ownership of our land spills out onto politics from a very grounded place—first, we are environmentalists.

We support the preservation of the county's agricultural land and work to slow development, we are vehemently opposed to nuclear power and offshore oil drilling, we monitor the county's timber industry for proper timber harvest techniques. We reforest our own land and clear clogged streams to allow salmon and steelhead to return to their native habitats. We boycotted a planned formaldehyde plant, we discourage all 'dirty' industry. We were the first county in the nation to ban the aerial spraying of phenoxy herbicides. These chemicals were used to kill the hardwoods in order to allow the more commercially profitable softwood forests to grow more rapidly. They are composed of 2,4-D, 2,4,5-T and Dioxin (like their sister chemical Agent Orange), and have been proven to cause miscarriages, birth defects, and genetic mutations.

The herbicide ban, accomplished in 1979 by a citizens' initiative and special county election, felt like a major battle won against the large timber corporations—a major battle won for ourselves and our environment.

The herbicide ban followed closely the election of Norman de Vall, the first environmentalist candidate to the board of supervisors in 1978. The same election gave us Joe Allen, a district attorney sympathetic to the absurdity of victimless crime and in favor of legalization of marijuana.

Enjoying a flush of success, three more candidates ran for supervisor on an environmental platform in the 1980 election. One, Dan Hamburg, won, one lost by just over 300 votes, and environmentalists were left 300 votes short of a majority on the county's board of supervisors.

Three years after United Stand united the community of illegal (non-code) cabin dwellers, a state suit brought legal development in Mendocino County to a halt until a general plan was written to more adequately deal with the future needs of this county.

Local citizens committees met weekly to form into a workable plan the needs of the present and the dreams of the future. New concepts of land ownership, proposed by the advisory committees include ways of increasing population density in specific areas while leaving large tracts of agricultural land in production, and allowing groups to share water systems, gardens and group living spaces. These innovations grew out of direct experience of living on the land. The economics of land ownership is a hot issue here as land prices skyrocket, and many see the concept of sharing land as the wave of the future. But these land use concepts are very different from the one house per parcel concept which has been the only acceptable method of land division for years. They are new, and considered radical by land-use planners.

The struggle between the needs of landowners and real estate developers, environmentalists and big business and the concepts of the county's planning experts are now being fought in the political arena—and much of the fight is bitter. The plan has been three years in the making, deadlines for completion set by the state creep closer, and the plan seems to take one step backward for every two steps forward. Until it is complete, there can be no further development in the county. Many live in a vacuum, not knowing if they will be able to divide the land as they thought they could when they bought it. Many love the stability the moratorium on growth has given their communities. Others resent those who continue to build illegally, and a new wave of red-tagging is sweeping the county, this time with fines of up to $250 per day for not complying with the orders of the building department.

There is a third struggle that divides the county—the issue of marijuana growing. Marijuana was listed as the county's second largest agricultural product by Ted Erickson, the agricultural commissioner, in 1979. Marijuana growers are linked by the political right with illegal cabin dwellers—they say illegal profits make our lifestyle easy, give us the means to finance political campaigns and the time to sit in the courthouse monitoring the supervisors and their actions.

At first, marijuana was grown for the joy of growing it: we planted the seeds we found at the bottom of our bags of acapulco gold, and plants flourished among the tomatoes in our gardens.

Soon we realized that we could grow enough so that we didn't have to buy acapulco gold any more: one more step toward a self sufficient homestead. Marijuana grew so well in our gardens that we had enough to give away to our friends. It was years before we thought of selling it.

It was the United States' spraying of paraquat on Mexico's fields that increased the demand for our new product; the discovery of sinsemilla (the cultivation of females in the absense of males to increase potent resins) that increased its value. Now, the weed brings more income into the county than lumber or grapes, previously the major agricultural products.

Untaxed—and illegal—income. Now marijuana growers are hunted. The local narcs, subsidized by state and federal funding, talk of aerial spraying the county with paraquat. Planes fly low over our tomatoes all summer long, peer into our canyons, photograph our orchards. When the narcs do make an arrest, they quadruple the plant count to bolster their reputations and the grower's sentence.

So now, even committed growers plant less, only the most choice, previously sexed female plants. They grow them deeper in the redwood and fir forests, water on systems separate from their homes, plant more sparsely to escape detection.

And they grow them bigger. The same agricultural techniques shared in the seventies to grow enough tomatoes for canning, and spaghetti sauce, and salsa for all winter are now applied to the most loved, and most economically valued crop. Because the plants are loved, are treated well, they grow big and strong—some 20 feet high and sometimes two pounds heavy.

But there are few people getting rich on marijuana. Most grow it part time, using a few plants to supplement their homestead incomes. They invest their profits in ponds, erosion control projects, more fruit trees, solar heating systems for their homes, working vehicles. And yes, in environmental and human rights issues, in political campaigns, both local, national and international.

Income from marijuana sales (most of which are out of county) returns home to support much of the rest of the community. Local merchants know this: they are angered with the increasing marijuana confiscations and arrests: they know that every pound of pot burned is $1500-2,000 out of their pockets. And merchants like to do business with marijuana growers; they find them considerate, they pay cash, and they always pay on time.

Still, most citizens of this very poor county, which has a serious housing shortage, a consistent two digit unemployment rate and upwards of 20% of its population on some form of public assistance, feel the need to continue to prosecute those who are attempting to provide for themselves, without the county's help.

We came here originally, not to be rich or famous, but because we thought this county was special. We have been joined by others who seem to agree. Land values have gone from $300 to $2,000 per acre in the past ten years. The population of the county has increased by 15%. The county newspaper wins national awards for muckraking, and our theatre group has been said to have more acting talent than the entire town of Seattle.

Perhaps this place is special because we are rooted here, we are landed, we care more about the products of our actions. We work to become good at what we do. For me, it is that same lack of a middleman I discovered nine years ago. A small community, direct contact, direct results. Constant feedback on the instinctive knowledge that everything is connected.

Perhaps I have made life in Mendocino County sound too idyllic, too dramatic. But although we have graded and gravelled our roads, we still do get stuck in the mud in the winter, we run out of firewood when the kids have the flu, we are watching the best of our bottom land rapidly turn into shopping centers. Some say the county is worse off than ever. It is growing too fast despite the moratorium. Although Anon Forrest, a United Stand representative went to Sacramento by request of the governor to help rewrite state housing law, our houses are still being red tagged, there is no general plan and the best of the agricultural land is disappearing under asphalt.

But those people who met at the first United Stand meeting are still frequent visitors to the courthouse, they now sit in the supervisors chambers with old timers who have learned to share their values, and with those who live in town and aspire to own land some day. You can't see the divisions in the community as well any more. All speak their minds on decisions being made by the local government.

There is a real sense of all being in this together. It is like when you buy land next to someone and you intend to stay there. You need to make friends with your neighbors, for they will be an integral part of your life for the rest of your life, and you hope, for your children's after that. Your roots will grow together.

Here in Mendocino County, our roots are growing together. Making us stronger. Little can shake us from the path we have chosen.

Hummingbird

My friend
Brad Ensminger
is in New York.
He's like
a hummingbird.

I thought I
saw him once
on a flower
in the Japanese
Tea Garden
but I'm not sure
it was
him.

—*Chris King*

WHAT THE BAY WAS LIKE

Malcolm Margolin

Beginning in 1769 small bands of Spanish soldiers and Franciscan fathers pushed north from Mexico to explore northern California and establish forts and missions. What kind of land and what kind of people did they find here? From their diaries and journals, and from the travelogues written by seacaptains who later visited the San Francisco Bay Area, we catch glimpses into an extraordinary world—a world that is as surprisingly unfamiliar to modern residents as it was to the earliest explorers.

In 1776 a Franciscan father stood upon a "very high and perpendicular cliff" (what is now called Fort Point in San Francisco) overlooking the Golden Gate, and he described what lay before him:

> We saw the spouting of whales, a shoal of dolphins or tunny fish, sea otters, and sea lions. This place and its vicinity has abundant pasturage, plenty of firewood, and fine water, all good advantages for establishing here the presidio or fort which is planned. It lacks only timber, for there is not a tree on all the hills, though the oaks and other trees along the road are not very far away. The soldiers chased some deer, of which we saw many today, but got none of them. We also found antlers of the large elk which are so very plentiful on the other side of the estuary.

Herds of elk were found throughout the grasslands. "Monsters with tremendous horns," as Francisco Palou, one of the early missionaries, called them. These were the Roosevelt elk, now extinct in the Bay Area, but then in such numbers that Palou and others compared them to "great herds of cattle feeding on the plains."

Packs of wolves were seen hunting the elk, antelope, deer, rabbits, and other game. Bald eagles and giant condors glided through the air. Among the larger predators were mountain lions, bobcats, and coyotes—now seen only rarely, but then far more common and visible—and of course the grizzly bear. "He was horrible, fierce, large, and fat," wrote Father Font. These enormous bears were found everywhere, feeding on berries, lumbering along the beaches, gathering beneath oak trees during the acorn season, and stationed along every stream and creek during the annual runs of salmon and steelhead.

Indeed it is impossible now to estimate how many thousands of bears lived in the Bay Area. Early Spanish settlers captured them readily for their famous bear-and-bull fights. Ranchers shot them by the dozen to protect herds of cattle and sheep which were soon introduced to the area. And early California chose the ever-present grizzly as the emblem for its flag and its statehood. The histories of many townships tell how bears collected in troops around the villages' slaughterhouses and sometimes wandered out onto the main streets of the towns to terrorize the inhabitants. Yet today there is not a single, wild grizzly bear left in all of California.

The ocean and the unspoiled bays of San Francisco and Monterey also presented aspects of almost heartbreaking beauty and plenty. There were mussels, clams, oysters, abalones, sea-birds, and sea-otters in profusion. Sealions covered the rocks at the entrance to San Francisco Bay, and were so thick in Monterey Bay that to Father Crespi they looked "like a pavement."

Long, wavering lines of pelicans threaded the air. Gulls, cormorants, and other shore birds rose, wheeled, and screeched at the approach of a human. And rocky islands like Alcatraz (which means *pelican* in Spanish) were covered with the droppings and nests of great colonies of birds.

Before the days of the 19th century whaling fleets, whales were often spotted within San Francisco Bay and Monterey Bay. LaPerouse writes: "It is impossible to conceive of the number of whales with which we were surrounded, or their familiarity; they every half minute spouted within a half a pistol shot of the ships and made a prodigious stench in the air." And several early travellers comment, quite independently of each other, on what must have been a common sight: a whale would get washed up on shore, and grizzly bears—or in many cases Indians—would stream down the beach to feast on its remains.

Whales and grizzly bears on the beaches certainly strikes us as strange, but then the whole of San Francisco Bay presented an entirely different aspect from what we have come to know. Rivers and streams emptying into the Bay often fanned out into estuaries which supported extensive tule (cattail) swamps. And in these days before landfill, the low salty margins of the Bay were surrounded by vast pickleweed and cordgrass marshes. Cordgrass provided what many biologists consider to be the richest wildlife habitat in all North America.

Inland from the Bay the land was surprisingly moist. Eventually a growing population would dig increasingly deeper wells and pump out more and more of the groundwater for its needs. But in the early days the water table was much closer to the surface, and the first settlers who dug wells regularly reported striking clear, fresh water within a few feet.

And water was virtually everywhere. Explorers suffered far more from mosquitoes, spongy earth, and hard-to-ford rivers than they did from thirst—even in the heat of summer. Places that are now dry were then described as having springs, brooks, ponds—even fairly large lakes. All the major rivers—the Carmel, Salinas, Pajaro, Coyote Creek, and Alameda—as well as many minor streams, spread out each winter and spring to form wide, marshy valleys.

Today only Suisun Marsh and a few other smaller areas give a hint of the extraordinary bird and animal life that the fresh and salt water swamps of the area once supported. Ducks were then so thick that one early hunter tells how "several were frequently killed in one shot." The channels that crisscrossed many of the swamps—channels so labyrinthian that Kotzebue wished he had a good pilot to thread his way through them—were filled with muskrats in fresh water, sea-otters in salt water, and everywhere with thousands and thousands of herons, curlews, sandpipers, and other shore birds.

The geese that wintered in the Bay Area were "uncountable," according to Father Crespi. Frederick Beechey, an English ship's captain, describes the area between San Francisco and San Mateo in 1826 where the number of geese "would hardly be credited by any one who had not seen them covering whole acres of ground, or rising in myriads with a clang that may be heard a considerable distance."

An early settler of the Bay Area, George C. Yount, describes the hills around Benicia in the 1830's:

> The deer, antelope, and noble elk . . . were numerous beyond all parallel. In herds of many hundreds they might be met, so tame that they would hardly move to open the way for the traveller to pass. They were seen lying, or grazing, in immense herds, on the sunny side of every hill, and their young, like lambs were frolicking in all directions. The wild geese and every species of waterfowl darkened the surface of every bay and firth, and upon the land, in flocks of millions, they wandered in quest of insects and cropping the wild oats which grew there in the richest abundance. When disturbed, they arose to fly; the sound of their wings was like that of distant thunder. The rivers were literally crowded with salmon. It was a land of plenty, and such a climate as no other land can boast of.

The early Bay Area explorers found here an immensely varied and plentiful wildlife population—far richer than we or our descendents will ever see again. The elk, antelope, wolves, bald eagles, condors, and grizzly bears are gone forever. And the animals that remain are not only greatly reduced in number, but in the last 200 years they have drastically changed their character as well. For as one reads the old journals and diaries, one is struck by the observation—surprising even to the early explorers—that the Bay Area animals were relatively unafraid of human beings.

Traveller after traveller commented on this. Foxes and mountain lions, which still survive in the Bay Area but are now quite secretive, were then prominent and visible. Quail were "so tame that they would often not start from a stone directed at them." Coyotes "prowl about in the most daring manner," according to Beechey, and in general "animals seem to have lost their fear and become familiar with man."

Kotzebue tells of bringing a crew of Aleutian Eskimos with him to the Bay Area to hunt sea-otters. "They had never seen game in such abundance," he notes. "And being passionately fond of the chase they fired away without ceasing." Then, one man made the mistake of hurling a javelin at a pelican. "The rest of the flock took this so ill, that they attacked the murderer and beat him severely with their wings before other hunters could come to his assistance."

It is obvious from these early reports that the animal world must have been far closer to the Indians of the Bay Area than it is to us. But this closeness was not without serious drawbacks. Grizzly bears, which in our own time have learned to keep clear of humans, were much more of a threat to a people armed only with bows and arrows. Jose Longinos Martinez mentions that during his short visit to California in 1792 he saw the bodies of "two men who had been killed by this ferocious beast." And Father Font also mentions seeing several Indians "badly scarred by the bites and scratches of these animals."

But suddenly everything changed. Into this land of plenty, where animals were familiar and relatively unafraid of people, arrived the European and the rifle. Within a few generations some animals and birds were totally exterminated, while others survived by learning to keep their distance.

Today we are the heirs of that distance, and we take it entirely for granted that animals are naturally secretive and afraid of our presence. But for the Indians who lived here before us this was simply not the case.

The Ohlones, like other hunting people, depended upon animals for food and skins. As hunters they had an intense interest in animals, and they had an extraordinarily intimate knowledge of their behavior. A large portion of a man's life was spent learning the ways of animals.

But their intimate knowledge of animals did not lead to conquest, nor did their familiarity breed contempt. The Ohlones thought deeply about the animals they mingled with; and like other hunting people, they held animals in great respect, fear, and even religious awe. They lived in a world where people were few and animals were many, where the bow and arrow were the height of technology, where a deer who was not approached in the proper manner could easily escape and a bear might conceivably attack—indeed, they lived in a world where the animal kingdom had not yet fallen under the dictatorship of the human race and where (how difficult it is for us to fully grasp the implications of this) people were not yet the undisputed masters of all creation.

If He Can Make Her So

One of the bad things about living is
To wake at dawn with dryness in the throat
That is half-way to choking, and to know
The dust is blowing: to wake and lie there thinking
Of the long quiet years before we came here
And violated earth's protecting girdle.

Hunching its wide shoulders, the storm comes striding
Across eroded mesas, through the orchards,
Into our doors and windows, and our thoughts.

The living roots enfold the soil, the soil
The living roots—between them is forever
The secret ritual of their nourishment.
Without the roots the earth must blow away,
And out of earth roots wither.

Nobody knows what the soil is, except
That it is something working towards a balance,
Something that balances itself with death
As well as life, and needs long years to do so.
If the wide earth has anywhere done better
Because of men, be sure they were good men,
Each of whom tended his own bit of ground
Humbly, and went down into it at last
His heart already changed to a rich compost.

Perhaps it takes an Indian not to harm
The earth he owns. It may be it takes praying.
Or else it takes a man out of the future—
I think that he, that future man, will see
That earth is truly part of his own being
If he can make her so, as his thoughts are,
And instincts too, if he can make them so.

—*Haniel Long, 1945*

The Hermit

A HERMIT CAME DOWN FROM HIS MOUNTAIN

THE HERMIT LOVED HIS MOUNTAIN
THE HERMIT COULD STAND ON THE TOP OF HIS MOUNTAIN
AMONG THE STARS
AND BREATHE THE RHYTHMS OF THE UNIVERSE

Kristin Wetterhahn

BUT THE MOUNTAIN GOT LONELY

A HERMIT CAME DOWN FROM HIS MOUNTAIN
    TO BE WITH HIS PEOPLE

HE DID NOT LIKE WHAT HE FOUND IN THE CITY
HE SAW PEOPLE CAUGHT IN THE ILLUSIONS OF POWER AND MONEY
        IF THEY HAD LIVED IN THE MOUNTAINS
    THEY WOULD SEE HOW FOOLISH THIS WAS

HE SAW PAIN AND SUFFERING

HE SAW MANY PEOPLE LIVED IN POVERTY
    AND MOST WORKED HARD FOR THE BENEFIT OF FEW

HE SAW PEOPLE FOULED THE LAND
      HAD LITTLE RESPECT FOR IT

HE FOUND EVEN IN THE CITY HE COULD BE LONELY

OH HOW HE LONGED FOR HIS MOUNTAIN

THE MOUNTAIN AIR CLEARED THE INSANITY OF THE CITY
AND THE HERMIT LOOKED BACK DOWN UPON IT
        AND REMEMBERED
NOT SO MUCH THE PAIN AND SUFFERING
    AS THE HUMOR THAT GOT PEOPLE THROUGH IT

HE REMEMBERED PEOPLE STRUGGLING TO BRING FORTH
    THE BEAUTY WITHIN THEMSELVES

HE REMEMBERED PEOPLE STRUGGLING TO CREATE THE WORLD
    A BETTER PLACE FOR ALL PEOPLE

AND ALTHOUGH HE WAS NOT SURE THE STRUGGLE
    WAS RIGHT-HEADED
WAS NOT SURE WHAT WAS RIGHT-HEADED
AND WAS NOT SURE THE STRUGGLE
    WOULD PRODUCE THAT MUCH

HE FOUND GREAT BEAUTY
    IN THE STRUGGLE ITSELF

MOST OF ALL HE REMEMBERED THE FEELINGS OF LOVE
    WHEN PEOPLE SHARED

A HERMIT CAME DOWN FROM HIS MOUNTAIN
    TO BE WITH HIS PEOPLE
    AND LEARN THE LESSONS OF THE CITY

—Tom Hile

Ed Buryn

MY SAN FRANCISCO

Stephanie Mills

I have lived in a lot of places around San Francisco Bay—Oakland, Berkeley, Point Richmond, Bolinas, and Tiburon, finding those towns to be too poor or too arty, too polluted, too spacy, or too rich. But San Francisco turns out to be just right—big enough, small enough, neighborly enough; radical, diverse, creative and cynical enough. It's a rightsize city, a city I delight in calling my home even though it is overrun by tourists, compromised by greedy developers and venal politicians, and starved by tax cutters.

When I see all the well-meaning rubes from Danville, Raleigh, and Cincinnati flooding into town, queuing up for the cable car, I want to send them home, to tell them to figure out how to see their own home towns as being as interesting as this. But I left my former home town—Phoenix—without having figured *that* one out myself. I shared the impulse, succumbed to San Francisco's charm, and burned my bridges.

San Francisco's founders were almost just passing through: opportunists, not sodbusters, and the tradition they established continues. Most San Franciscans are critical of the shape their town is assuming and nostalgic for the static little jewel of a city it never has been. Since the advent of the Yankees, this capitol of the west has housed get rich quick schemers. Earlier on, the Costanoans and Californios knew how to have a pleasant unambitious existence in this bountiful and temperate place. Their contentment scandalized the New England traders arriving here. And the Forty-Niners whose arrival turned a pleasant Mexican village (and all the redwoods in its environs) into a city overnight were so speedy and profligate that they left their ships behind them to sink into the murk of the waterfront like so many empty beer cans.

My living room overlooks the resting places of one of those ships, which became landfill and is now a hundred yards from the waterline, waiting to resurface in the next quake. She was briefly exhumed during the construction of Levi Strauss' world headquarters, but Levi's (founded during the gold rush, incidentally) was in too much of a hurry to allow a full scale archaeological study of the site, and so there her timbers rest, reburied to await the next orgy of razing and reconstruction.

The razing could be done by nature or by man, given the city's location. We live right at the spot where the Pacific floor slides under the continent, churning up an unstable geology. Because of this, we share a sporting contempt for death, a willingness to risk chaotic demise for the sake of living with all this unsteady beauty.

San Francisco Bay is a more pleasant feature of the landscape. Inhaling fog is the worst thing it does. The best has been making San Francisco a multicultural port city. Because we face Asia, we anglos have to stretch our consciousness to grasp that of the East, and are the wiser for it. It's only a modest assertion that physically and historically, San Francisco is uniquely blessed.

• • • •

When I first came here as a teenage tourist I was seduced by Chinatown, maybe only Grant Avenue, the tawdrily painted face San Francico Chinese grin at visitors from. But this one among scores of ethnic neighborhoods was everything my hometown wasn't: aged, ornate, other; decorated, redolent, kaleidoscopic; full of people who weren't Mormons. I had had a glimpse of a culture older than mine and I had to return to be with it. I hungered to be in a real city.

Later, as an undergraduate at Mills College in Oakland, I often returned to San Francisco by bus. At that time, women still wore gloves and hats here, which gave me an opportunity to play dress up, wearing clothes that would have stifled me in Phoenix.

For a long time, my San Francisco was mainly touristland, Union Square, with its colorful Mafia flower vendors, and Nob Hill, where there is nothing to do except be reverent in Grace Cathedral or sit prettily in Huntington Square beaming at children. Then, the city was a theatre to pose and behold in. I just watched people, imputing plots to their lives, pitying them, suspecting that romance or tragedy lay behind feigned indifference, and only rarely failing to script the scene. Once in a great while that writing went silent and I fell to wondering just why somebody happened to be where they were when they were; what fate assorts people on a certain street at a single moment? I never found an answer.

San Francisco was a nice place to visit and I did want to live here and become part of the assortment, but it took me about ten years to make that circuit of Bay cities to do that. Living here, I still like to travel on foot, reading the town line for line, observing my neighborhood most closely of all. Here, people do imaginative deeds which add visual interest; like the jade trees gracing the windows of so many Chinese businesses reminding us to revere nature. They do whimsical things, like the welded iron Uncle Sam's head on a red white and blue striped pike outside a barber's on Columbus avenue; make offhand jokes, like a preppy fisherman's hat placed on an anxious iron hound in front of a tony apartment building. There are artistic statements like the squad of black and white cutouts of Montgomery Clift which appeared pasted just at the edges of walls all over North Beach. These lifesize photographs of a dapper young man suavely smoking and watchful were everywhere one day and then gradually they disappeared—were defaced or removed.

• • • •

These days, I live in a Telegraph Hill apartment with a magnificent view of the Bay Bridge. The Hill used to be a Bohemian place—a slum that saved itself from being saved, by the dynamite firefighters after the '06 quake, by dint of being worthless. There are still certain inconveniences associated with living here—a terrific amount of stair climbing and a parking problem which makes us feel a little daring, but on the whole, the neighborhood is cushy and likely to become more so, as Levi Strauss' world headquarters burgeons at the foot of the hill, threatening an onslaught of gentry richer than we, better able to afford higher rents.

The last real Bohemian—a squatter— was driven away ("for his own good") from his earthquake research lab shanty near the Greenwich steps by the construction of a multi-story parking lot. There are a few writers, but they're fairly well-heeled. There's a very diligent pianist down the steps a way who adds musical pleasure to the pleasure of walking among the Filbert Steps roses and fuchsias with his practicing. There was a painter up the hill in the pink stucco with the purple bougainvillea. She was fey, wore pigtails, and spent a lot of time with her daughter. Once she got a commission to paint the Ringling Brothers' Circus clown car, which she did out on the street. She covered a little Toyota wagon with Chagall-like clowns in green and red and yellow. That was the most artistic happening I've seen around here to date. So it's all very genteel, and I'm sure our generation of careerist swinging singles managed to up the rents and drive out a previous group of struggling young postwar couples just as the executive immigrants from Piedmont and Lafayette may displace us.

Meanwhile, our landlady, caught in the inexorable logic of inflation and interest rates, is eager to sell the building, as her investment in it is returning less than it would in a bank. I try to be nice to her realtor on occasions when I'm here and he shows his prospects through, but it's a struggle. I've been evicted before and I hate it.

• • • •

Before moving to the east side of Telegraph Hill, I lived down on Stockton Street, in the heart of the dwindling Italian neighborhood, which is a good place, rich with Latin character.

A couple of Sundays ago, walking home through the old neighborhood from a swim, I saw a half dozen little girls dressed as brides, surrounded by their parents and grandparents as they got into cars or simply walked down the hill to SS. Peter and Paul for their first communion. So dazzling and lacy in the sunlight, these children were taking their first steps into adulthood, espousing Christ, shod in white patent Mary Janes. New life for the old church which is the obstinate heart of their community.

In the railroad flat opposite ours on the top floor of the Stockton Street residence lived the widow of the drayman who had built the place in the twenties. She was an old Ligurian crone of about 96, and her life had gotten pretty spooky after all those years. Because she spoke little English, she was very dependent on her daughter, who did the landlady job, and who disliked our noisy lifestyle enough that she eventually invited us out.

It was an old building, and fragile. It was property, and sacred. The landlady had a million rules for living there, all of them intended to forestall her ever having to spend any money on repairs. Living there was a lab course in the culture of aging Italian immigrants and taught me that the Young American way of life, with loud parties and much coming and going, is not universally liked.

Shortly before we left the place, there was a minor catastrophy which brought us into closer contact with our neighbors, as catastrophes often do. It was Labor Day weekend and the old lady's daughter was out of town. I was left in the middle of a deadline, very fucked up and paranoid behind some cocaine on which I'd stupidly squandered my nonexistent funds. Doubtless, too, I was hung over, because we had entertained the night before. This fact made us the prime suspects in the crime of breaking the sewer pipe, which had happened that morning. The old lady hammered on our door bright and early. She was distraught, hysterically crying, "You breaka da pipie! It a no workie! It a no workie!" I couldn't cope, and fled to my typewriter, angry that we were being blamed for this.

My partner, Jamie, listened sympathetically however, and went to inspect the damages. He called his friend Gundars over from Berkeley. Gundars arrived clad in overalls, carrying a massive pipe wrench—not that he knew anything about plumbing but the getup had a calming effect like a doctor's white coat. Gundars and Jamie messed around with the pipe for several hours to no avail.

Part of this messing around required their going into the garage on the south side of the building where they discovered untouched racks of crusted wine bottles and huge casks full of a violent old Dago red, home made by the old lady's long gone husband, perhaps during Prohibition. In spite of the fact that they hadn't fixed the pipe, to show her gratitude, she offered a bottle each to Gundars and Jamie, saying she had never tasted it herself. Gundars got a bottle from 1927, Jamie one from 1934. Months later we cracked ours open to celebrate our move from Little Italy, and it was as fierce and murky as its maker's old wife, our first and last communion with her.

The old neighborhood keeps losing a little of its backbone because the pious old world characters are dying off and taking their virtues with them. A delicate, ailing seamstress used to have her shop downtown Stockton Street across from the Liguria Bakery. It was a pretty little place painted robin's egg blue outside and white gone ivory with age inside. She was short and wistful, a self-reliant spinster with frizzled hair. On her powdery cheeks were two bright spots made not by rouge but tuberculosis, I guessed. I went to her once to have a jacket shortened, and though our conversation was limited by my ignorance of Italian, we communicated well enough on the question of sleeve length. She coughed, and said she had been sick for a long time now. I needed to try on the clothes for her to check the fit, so she led me through a curtained doorway into her apartment behind the shop. It was a perfect nun's cell, with a narrow bed neatly made, a tiny stove with a few exactly placed pieces of fiesta ware on a shelf above it, a painted wooden table and chair, a gleaming empty sink, and a religious calendar on the wall. That was all. Her home was austerely dignified, making most others seem frivolous, excessive.

A few days later, I collected my jacket from her. I never saw her again. Some months afterward a tax accountant took over her shop, leaving me to speculate on her chaste and perhaps lonely death.

• • • •

The old Italians, the Chinese, the gold rush crazies—all of them laid down their tracks in this city. Everyone takes turns exploiting the most recently arrived, although it takes great cleverness to exploit the Yankees who wrote the book. Asians and Arabs are that clever. The squeeze is on, housing is harder to find, the poor are driven from their sunny old neighborhoods by monstrosities like Yerba Buena, which will deepen our dependence on tourism; or by the upwardly mobile former immigrants. But that churning is what we come to cities to be part of. Growing or shrinking, San Francisco has always been dynamic and the era whence emanates the columnists' nostalgia was a flicker.

For all the opportunism in the city's past, we must still be Costanoans at heart. Enjoying the pleasures of San Francisco, and they are numerous—even Parisians admire San Francisco—is not without its price. The cosmic price is set by the San Andreas fault; the existential price is that this isn't where it's at professionally for much of anything. So because American cities specialize, the decision to put down roots in San Francisco is primarily hedonistic. We avoid the adrenal rushes and can only play king of the hill in a smallish way. Maybe that's why San Franciscans seem so charming to visitors.

San Francisco isn't the set Karl Malden and I took it to be. Nor is it merely a place to work, although downtown is to an excessive degree—a desert by night. It is a place for people to live in, to build cultural or generational niches in, to supply the human richness of detail. We are watchers and watched, neighbors and friends: inhabitants.

Arthur Okamura

Stake in My Heart

The cash register resounds behind the band
and the sax man splits a lung
trying to get back to Africa
but no luck, it's Los Angeles

nevertheless there's moonlight
fuming over low clouds and the crush
in here intensifies as fast talkers
jockey for position—
                    my lewd aunt

prompts me from her apartment, flutelike/
like a whistle only curs can hear,
with a sense of fun behind her bizness
front, wanting
to get rich & live
to see it on film.

All the dog biscuits broken
in honor of the stray
crumble slightly at the sight
of his fangs which rabidly
defy domestication, intent

on returning to the sharp
heartache, that flame
beaten repeatedly till tempered
& found fierce, resilient.

—Stephen Kessler

Deeper into the Mundane

Gray city dawn,
I woke from a dream
of small town, school & lost love
in a west side apartment
in the fold-out sofa
of a working girl gone to the yoke,
her roommate next door
dozing and my tears
dripping in the shower.

Hotcakes & eggs & coffee
with a side of carbon monoxide
excite me on the wide boulevard
while my Italian fuel pump is being fixed.

The closed shops, the shoe repairman
moping in his open one, the overcast
reflect the slow homesickness
suffered by those whose
heartaches were demolished for moderner
emotions—

    "Something's moving," sez
Einstein from the side of a city bus—

it is the bowels of thousands
one September Wednesday,
a lake in far mountains losing
five gallons each time
a toilet flushes in town.

We are so many and we
need so much, nothing
we touch is unchanged—

reclusive insight useless to seize the escaping

—Stephen Kessler

SAUSALITO WATERFRONT

Phil Frank

MARIN COUNTY, WHICH HAS OF LATE BECOME A MECCA FOR THE NOUVEAU RICHE AND THE SELF GRATIFIERS HAS ALSO ATTRACTED LARGE SCALE OFFICE DEVELOPMENT... THIS HAS BEEN MOST NOTICEABLE IN THE SMALL TOWN OF **SAUSALITO**.. WHERE ITS WATERFRONT, TRADITIONALLY A FISHING AND BOATBUILDING AREA IS UNDER SEIGE... WE NOW TURN YOU OVER TO OUR HOST:

MUCH ATTENTION OF LATE HAS BEEN DEVOTED TO THE MANY CHANGES ALONG THE SAUSALITO WATERFRONT. TO MANY PEOPLE ALONG THE SHORELINE THIS AREA IS A WORKING **WATERFRONT ZONE**.. FOR DEVELOPERS IT IS CONSIDERED AN **OFFICE PARK ZONE** BUT WHICH, FOR OUR PURPOSES TODAY WE'LL CALL.... THE **TWILIGHT ZONE**..

..FOR THERE ARE UNEXPLAINED PHENOMENA AFOOT OF LATE...

GUIDO MILANO THIRD GENERATION FISHERMAN:

STRANGE THINGS ARE HAPPENING! THE OTHER DAY I'M DOWN ON THE FISH PIER MENDING NETS WHEN THIS GUY APPEARS OUT OF NOWHERE.

I...MEAN...YOU..NO HARM...CAN... YOU... HELP,.. ME?... I.. WANT.. ..TO.. MAKE.. A.. XEROX. ...TAKE..ME.. TO.. YOUR,. COPIER..

..AND THEN HE WAS GONE..JUST LIKE THAT. I'VE WORKED HERE 25 YEARS. NEVER BEFORE DO I SEE SUCH A THING.

A FIGMENT OF GUIDO'S IMAGINATION, YOU SAY?.. ..AN ISOLATED EXAMPLE OF A WARP IN THE COSMOS ? POSSIBLE... BUT LET'S LOOK AGAIN..

PLACES 123

Night

Freeways of night;
soul voices, young
in rhythm/
in *sense*,

stolen from fools.

(this American night
captured in song
by black girls
in high heeled shoes)

Or *engines* of night.

"I'll treat ya, baby"

Or bus drivers of night,
drivers of night,
of night—

curbing & arteries
& cities & cars
& swollen night
enveloping homes
of the laughing
dead . . .

(the radios of night
across America
signalling the
streets to
breathe)

Neon of night &
silhouettes,
the breathing night,

the tears & joys
& night of nights.

Of one cement ribbon
& a heart.

Of eyes injected
with the ragged
horizon.

Of the last intelligible
spoken word.

—*F.A. Nettelbeck*

Kate Drew Wilkinson

THE SAUSALITO HOUSEBOAT COMMUNITY

Piro Caro, talking

I came to the Sausalito waterfront in 1951. What I found here was very compatible to me. The waterfront was thinly populated by people who thought of themselves as traditional artists, painters, and musicians. People interested in culture, in bringing out a magazine of literature or in making crafts like jewelry and pottery. Sausalito by then had ceased to be a very active fishing village, but there were still some fishermen and fish-buying docks. A number of the fishermen lived in their fishing boats tied to the shore or anchored out.

But mostly what I was first acquainted with were painters and designers. The first year I was here, we decided to put on an Art Festival on what was then called Shell Beach in Sausalito. It was a very attractive and very good art show. Harry Partch produced his opera on Oedipus in full regalia. The actors wore traditional Greek masks. The actors were excellent troupers, thoroughly skilled and experienced in what they were doing, and the performance was a marvel. Harry Partch at the time was known to very few. He's since become an extremely important avante garde musician introducing his own kind of musical instruments. He made all of the musical instruments he ever used. They were new and they were different. Then there was Jean "Jonko" Varda, a painter who came out of the Paris school of cubism at the time of Picasso and Braque and those early painters of the early part of the century. He lived on a ferryboat, had a huge, very handsome sailboat, and was a great hand at giving elaborate parties. Jonko was famous as a raconteur and party giver.

Then there was a slow transition away from the high art of people like Varda and Partch to crafts—pottery work, jewelry making, weaving, that sort of stuff. In later years that Art Festival which we began in 1952 became a commonplace annual festival in which one sees some paintings but mostly pottery. Another thing that followed the high art period was the appearance of rock musicians. We had people like the Grateful Dead and Quicksilver Messenger Service, as well as other bands whose names, unfortunately, I don't remember. We had people like Lord Buckley, an enormously interesting circus poet. A man with the manner and voice of a circus barker making poetry out of jive. The extraordinary poems that he boomed were altogether tender. A circus barker telling in jive the life of Jesus.

This was still in the middle and late 50's. People who would perhaps allow themselves to be called beatniks were now around the waterfront. But, they really weren't beatniks. They were people who wished to publish rather staid magazines. For example, there had opened at that time under Herb Beckner, the Tides Bookstore. It attracted a certain kind of poet. Those people published a magazine called *Contact*. I think they got out one or two numbers.

Next was the appearance of the hippies. The highways of the United States became populated by young people running away from home, anywhere from thirteen to twenty-two years old. The domestic life of their parents was unacceptable to them and they were out on the highways looking for how to make a new world for themselves. They made the Haight-Ashbury their home and a certain amount of them fell to us.

What these young people found, as they came along highway 101 and looked at the houseboat community on the Sausalito Waterfront, was that we would receive them. Nobody cold-shouldered them, nobody would say no to them. It was absolutely open. The man who owned this waterfront, Donlon Arques, didn't ask anybody anything. He didn't ask them to come to him and say "May I?" He didn't ask them for any rent money and he never ousted anybody. In fact, he didn't much come about. He ran at that time a boat yard in downtown Sausalito, the Johnson Street yard which he later sold.

Anyway, these hippie people, these highway travelers, would come and say, "hey, can I build a houseboat?" and everybody would say, "Sure, go ahead." And we'd move over a little bit and give them a place on the beach and tell them where to get the salvage materials to build it. What had happened was that in the 1940's the Sausalito waterfront housed a shipyard built by Bechtel. It turned out one Liberty ship a day with 20,000 people working in the yard. When World War II was over the shipyard was abandoned, leaving behind mountains of timber, plywood, and landing craft hulls. All sorts of stuff out of which, of course, you build houseboats. These materials were still altogether available and altogether free when the hippies arrived. When a young man showed up and wanted to make a houseboat, we put out tools for him. We thought, if we didn't put them out, he'd steal them. So we might as well let him use them to begin with. Besides which, we knew that when we first came, we stole our materials and we stole our tools. For we had no money either. And so houseboat after houseboat was built for little or nothing and the community grew in this place during the 1960's.

It also, of course, was a haven for drug users and junkies. That was disapproved of outside of the houseboat community, and even somewhat inside the houseboat community. But, in fact, nobody interfered with the use of drugs. If people used drugs, we saw that they did nothing—the junkies never used violence. It was clear that these were very peaceful people. It was thought that they would rob everybody—for they needed the money to buy their stuff and they would have to steal to get it. Well, I don't remember that they stole. I do know they never stole from me. They did not steal from my neighbors. Did they steal from the solid citizens in Sausalito? I don't know. I haven't any idea. I do know they made very peaceful and sweet neighbors.

One thing seemed to be clear; we were that very curious thing which can't be seen—something called a community. I can't even describe a community. I take it that a community is an organism, like one's own body. You don't know what your toes are doing, nor how you send messages to your fingers. Yet, there they are, part of what you need to use. And so, generally that was the way it was for all of us on the waterfront. We were always at the service of each other. We were clearly interdependent. The young people who came and started to live here found a rich world. Soon any boy that came here found a girl. Soon that boy and girl had children. We all lived together and reared that strange anomaly in American culture, a community. Something not to be seen anywhere else within my recent experience. I was then in my 60's and marveled at how each of us knew of the talents, strengths, abilities and possessions that each other had and could call upon.

Many of us wished not to have to go out into the regular job market to earn our livings. So we, more and more, earned it among ourselves. I was a gardener and had to earn part of my living outside of the community. There were some within the community who went entirely outside of the community everyday to do whatever they did. Doctors, lawyers, engineers, teachers, philosophers. Some of them, like Alan Watts, were very luminous people.

So it began to be apparent that we were extraordinarily talented people charged with creative energy. There was a man, for example, who wanted to free the world of the strictures of the automobile. So he undertook to redesign automobiles. He worked out a method of making bodies out of fiberglass. He molded very handsome and beautifully colored automobile bodies. His impulse was to free automobiles from the formal, determined lines that Detroit imposed on them. This was just one "vagrant" person. This was his stuff. Other people were interested in alternative sources of energy. They experimented with solar heat, solar generators, and electric generators. Then, of course, all over the place could be seen wind generators trying to free us from the PG & E. Among us were people who cared a great deal about not using the Roman sewer. We thought that simply flushing into the toilet to give sewage a ride through the sewer pipe to downtown Sausalito, there to be dumped raw into the Bay, was nonsense. Consequently, there was a great deal of experimenting among us by people who wanted to make compost privies, the compost to be used for gardening. I, for example, have had a compost privie on my houseboat since 1967. It works perfectly well.

All over the place were people free of the need to give themselves nine to five to other people's affairs and work on their own. There appeared among us a great deal of vernacular architecture free of the building code. We built by necessity, by the beauty of necessity, houseboats that were very interesting and lovely. Christopher Roberts designed pleasure domes and called them houseboats. He built, for example, the one which is known among us as the Owl. He undertook to build another dome, a huge sculpture, 69 feet tall, that was to be in the form of a woman and child. It unfortunately caught on fire when it was almost completed and burnt to the decks. He also got four huge drydocks, bound them together with mud, and put them in the middle of Richardson Bay off Sausalito. He designed a city to be erected on them, but the City of Sausalito and the County of Marin forbade it. They took him to court, bankrupted him with lawyer's fees, and finally seized the drydocks. He gave up and went to New York.

It is clear, of course, that what is central to everything here is an aesthetic. We dearly love living here. It moves us. Nobody living here is less than sharply sensitive to the beauty of their environment. We speak of our morning light and our evening light, we speak of the views open to us. You can see from my windows, over my decks, San Francisco one way and Mt. Tamalpais the other. You can see huge vistas of water. The anchor-outs have the most idyllic, the most serene, physical situation imaginable. It's extremely beautiful to row out to, or row away from a houseboat that's anchored out. It's gentle and peaceful living out on the Bay for almost all of the year. But, there is a part of the year in which it is not, in which it is extremely startling. When the southeast wind comes in off the ocean, every boat anchored out is in great danger. It requires expert handling to keep from being driven off your moorings and into other boats or onto the shores. The southeast wind is a great menace to us. There is no breakwater that stops it. The prevailing wind, which is westerly here, is broken by the ridge of the hills behind Sausalito. So that it's only infrequently that even the fog comes down to us. I wish I could think of how to tell you of the aesthetic quality of living in this place. It's a tingling awareness to each of us.

Resistance and the Future of Low-Cost Housing

I've been speaking about two hundred and fifty houseboats, occupied for 25 years by people who pay nobody any rent or any taxes. While I've said that the most important factor in this houseboat community was aesthetic, the most apparent factor was economic. Not paying rent or taxes freed us, but that freedom had to be paid for. That freedom brought the hostility of the local governments, the City of Sausalito and the County of Marin. But their hostility entrenched our community. We had an outside enemy. We had a force outside of us trying to destroy us. So, for all of the time that I have lived on this waterfront, I've been involved in resisting the efforts by the City of Sausalito or the County of Marin to get rid of us. For 25 years we've resisted. They harassed us, they threatened us, but they never really were able to get anywhere. They charged us with violating the building code. But there was no building code for boats and the building code for houses on land simply had no application to us. So they wrote a building code which was obviously punitive and so outrageous as to be unenforceable. They wrote it declaring that we would have to get an occupancy permit if we wished to live in our homes. We could not get an occupancy permit, however, unless we met the building code that they had written. But the building code that they wrote was so dreadfully bad, they never tried to enforce it. They never required that we get an occupancy permit.

Instead they hit on a new device in 1969. They went to Don Arques, the landowner, and said they would give him a permit to build a marina in the space we occupied. He said, "Good, I'll take it," and told us to stop resisting. He told us that we had now got them just where he wanted them and that we had made him a half million dollars. But, what they planned was that the area we occupied should be turned into a fashionable houseboat marina. They gave Arques permits to build in 1970. But, it was 1977 before the development actually started. When they started to do it we looked at the plan of what they wanted to do and were horrified. They wanted to build four docks, 660 feet long, straight out into the bay and line houseboats up six feet apart from each other all along each dock like cattle. Who would want to live that way? We said that we thought the design for the marina was too ugly and ought to be forgotten. The developer, who had leased the waterfront from Arques, said he would revise it but he didn't. We quickly went to court asking the court to issue an injunction restraining the development of the marina until such time as a court of competence could determine the equity of our protest. We wanted first an environmental impact review before such buildings should happen. Second, we wanted the laws of the state and of the county to be enforced that called for the provision of low-cost housing, adequate to the needs of the community. We were the only low-cost housing in all of Marin County. And they were now proposing just to demolish it.

The court, after lengthy hearings, refused to issue such an injunction on the grounds that we came into court with unclean hands and seven years too late. We were told that we should have brought our suit to court when the permit was first issued in 1970. Instead, that judge issued a restraining order against *us*. He issued an injunction *for* the developers forbidding us to interfere with the police, the developer, the developer's workers or with the developer's agents; we couldn't interfere with their entering any of our homes at any time, or with their removing the lines that moor to our places, or with their moving our houseboats to any other place they wished.

This injunction, by the way, was remarkable in breaking our spirits. We now saw clearly that the judicial system as practiced in Marin County was not available to us. We were clearly put on notice that we did not have any standing before the law in the courts of Marin. It was a severe blow. For we didn't have very much else to go by. The tactic that the developers then went with was to employ Marin County sheriffs to terrify us. The police came in, time after time, in large numbers of 80-100. They spread out all over the place, tear gassed us, assaulted us with spikes and other outrageous weapons, and arrested us and took us to jail. They openly said that their purpose was to terrify us. To break the spirit of the resistance. But, we weren't doing anything to resist. All we ever did when the police came with their great numbers was to walk out and look at them. We got busted for that.

It's extremely important that there should exist somewhere an interesting outlet for people who cannot find and cannot afford the housing available in the regular world. There is, of course, an equally desperate need for young people to establish themselves in a world where they can find themselves competent to operate. Can they in San Francisco? Some can, some cannot. What happens in a community such as ours is that everybody finds his degree of competence and is secure and able. But, the need for low-cost living is universal, prevalent everywhere. There exists along San Francisco Bay hundreds and hundreds of miles of shore where such communities as ours could exist. The law, the basic law, is that all marshlands, tidelands, and shorelands are public trust lands established as such by English law since the time of Edward the Second of England. When the United States was formed as a nation it accepted this as the basic fact in America. As each state entered the Union it took over the protection of the waterways within its geographic limits. A curious thing happened in California. In 1869 the state legislature appointed a commission called "The Marshlands and Tidelands Commission" whose function was to protect the waterways of the state. It sold waterfront in San Francisco, Oakland, and Stockton in the 1870's for the erection of docks where ships could moor and unload, carry on commerce and trade. This was a very appropriate and proper thing for it to do. But it did more than that. It also sold the shorelines of places where no such thing would happen. For example, on the Sausalito waterfront 2-2/10 acre underwater lots were sold in San Francisco on Sunday mornings by the State of California. The state did not tell the purchasers that these were underwater lots. It was simply a real estate swindle. However, in the course of time those lots became valuable, too. Now, *that's* the water that we occupy. In August of 1979 the California State Supreme Court ruled that all previous verdicts of the courts of California in litigation affecting tidelands and public trust lands had been in error. It did not cancel the title of prior private ownership to these waters but it did reassert the prior right of the public to the *use* of those lands over their title owners.

We, however, think that it is purposeless for us to appeal to that court, or any other court in this state, for enforcement of this ruling. We don't think it's possible for us to get a non-political decision out of the courts. The politics of Marin County will determine any decision of the court, so we don't dare test it. And we can't afford to, besides. A legal aid society lawyer said he'll never again handle any litigation of the houseboat people. "It cannot succeed." He felt professionally impotent because of us. Well, let us not test, or let us, as the case may be, the use of waterfront land to which somebody or other has title. There still remain enormous stretches of waterfront to which nobody has title, stretches still under the protection of the state. If that public trust land, still publicly held and not privately claimed, were allowed by the local authorities to be occupied by houseboats, an enormous release of the housing shortage and housing costs could come about. Hundreds of miles of coastline could be used by houseboats, built for only a few hundred dollars apiece. But the city and county governments aren't really interested in responding to the urgency of people's needs. They traditionally are only interested in protecting property values and advancing development interests. If we built houseboat communities along public trust lands would we menace the interests of these people? I don't think so. But, I think that the local governments think so. They don't realize that low income houseboat communities are a creative alternative to public housing. Not welfare, but allowing people to build their own homes. What a thing to exercise the talents of a young, or even an older, person! To give them a feeling of competence, of place, of resoluteness, of being self-reliant.

I'm 80 years old, I'm blind, I've been somewhat paralyzed by a stroke. It's pretty hard for me to walk and get around. I can't see a nail to drive it into the wood. Yet, in this last year I built this houseboat. I didn't physically build it, two friends of mine did. But, I was necessary to the doing of it. It was my feeling of self-reliance that made it possible. Thousands of other people in need of a home could do likewise, if but given the opportunity.

Who Drive the Night

After midnight the trucks roll alone between Cleveland and Toledo
No cars, no cops, no cycles push the highway this time of night
Just trucks, 5 axle freighters peeling the stitches off the road
Trucks loaded with beef, sugar, liquid oxygen and chlorine gas
Trucks to Denver, to Kansas City, to Dallas packed with dog food
Appliances to Tijuana, sausages for the convention in Detroit
Hundreds of trucks, pulling in, pulling out, bumpered up for miles
Yellow lights, amber lights, roof bulbs blinking as they pass
Shouldering the rigs, the drivers weave an intricate coccoon
Meshing levers, valves and gears; human blood and diesel fuel
Hurtling along one vein, a vein of asphalt, shredding rubber
Bugs grilled and glassed like the petals of thrown bouquets
Back to back the trucks roll, the headlights galvanize the wind
Miles ahead the road forks, it fractures like the spokes of a wheel
Out across the night of a land that sleeps those who drive the day

—*Fruud Smith*

Just Past Black Shine

Do you hear the midsummer call of the wild blackberries?
A soft, seductive invitation.
A murmuring: full, ripe, ready.
Joyous song of a single juicy berry
as it slips down your parched throat.
Each day the call grows more insistent,
now a mosquito-buzz in your ear,
a plaintive fear of drying, unused, on the vine.
When I was younger, I picked only
blackberries shiny as new-washed faces;
even sweeter are those just past black shine
before the wrinkling and the shrinking.

—*Janet Carncross Chandler*

from:

MORDECAI OF MONTEREY

How Mordecai Took Ed Willits on an introductory Melanoia Spree

Keith Abbott

"This leads us into the problem created by our feelings of exuberance joy ecstasy satori, whatever. These are basically antisocial feelings."

—Philip Whalen

When Hilda called, Buck and Mordecai were at Buck's house having a large dinner. They had been too tired from logging to go back out to Carmel Valley and finish the moving job.

"It can wait," Buck had said, "I'm hungry. Besides, Hilda said the doc is in for a week long stay. There's no hurry."

"He was really upset when we started moving his things out, wasn't he?"

"Oh, he got his granny gland going when I dropped his desk drawer. Just looking in there you could see he liked his pencils arranged just so. Let's eat."

The phone rang. Buck answered it. "Yeah, oh Hilda. How are you . . . what? What he call me? A *wild* man? Me? Why that old fart's nothing but a walking granny gland. What? Granny gland, granny gland, you know, it's the gland that goes squeak and leaks when you get petty. Hmmm? Hell no, we didn't hurt his house! Never even came close! Aw, we dropped a pine and flipped one little bush up on . . . listen. I was raised in Tennessee and if *I* don't know how to drop a pine over . . . hmmm? I'm just having dinner. Sure, we'll drop by. Uh-huh, well you just have the gin and tonics ready. You heard me. We'll talk business then. Uh-huh, by the side of the pool. I'm going to need some cash in advance to get this tree job done right. You heard me. Okay."

An hour later Mordecai was wandering around Dr. Verog's house while Buck and Hilda discussed the financial arrangements over some gin and tonics out by the pool. The house was no longer so bare. Dr. Verog's own psychiatrist had recommended that the house in Carmel Valley be left as it was, so Hilda rented a bunch of furniture. She didn't want to add to her husband's trauma, she said.

Mordecai opened the door to Dr. Verog's study and found himself face to face with Ed Willits. He was Dr. Verog's assistant and he was busy making sure the study was exactly the same as the day GTM (Get The Money Movers) first made a beachhead on its propriety. Ed was a tall serious young man who always paused a half-second before saying even the most innocuous things. Mordecai was finding out a few things about Dr. Verog's brilliant history when Buck came in to tell Mordecai he wasn't needed any longer.

Mordecai introduced Ed to Buck. "Oh, you're his assistant," Buck said, eyeing him, "Mordecai why don't you tell Ed about your disease. Mordecai's got a rare mental disease."

Mordecai told Ed all about his melanoia, how he was always thinking that he was following someone, that there was a huge cosmic conspiracy to do good, and how someone was about to give him something while Buck stood behind Ed and pointed in mime that Ed and Mordecai should leave.

"I've never heard of that," Ed said.

"Oh it works," Buck chimed in, "in fact, it's probably working right now, isn't it, Mordecai? Melanoia needs the bright lights," pointing down toward Monterey, "doesn't it, Mordecai? Why don't you two hit the bright spots?"

"Well, I do feel like I should be following someone, and," Mordecai looked over Buck's shoulder at Hilda who was in a bikini by the pool and pouring the drinks, "I do feel that everything's going to work out fine."

"I've got the same feeling," Buck said, peeling off a fifty and handing it to Mordecai, "why don't you take Ed here out on an introductory Melanoia spree. You won't believe, Ed, just how powerful melanoia can be."

"Now wait a minute," Ed said, observing Mordecai, "you mean you don't feel any *compulsion* to do these things? Do you feel like you have a *choice*? Or do you *have* to do this . . . melanoia? Or is it a *conspiracy* to make you feel optimistic and you don't *really* want to feel . . . fine."

"You have to see this in *action*," Buck said, "do some field work tonight. Just observe the pateint here and see what you think." Buck began walking them toward the door. As they passed through the front room, Mordecai could see Hilda sitting on the patio, wearing a bikini. "I'd love to go but Hilda and I have to straighten out something."

So, in the interest of science, Mordecai and Ed drove down to a bar on Alvarado St. and Mordecai began to explain how the disease worked for him. He was just explaining how precisely systematic the melanoia was when a lumber salesman by the name of Sean LeBaron sat down by them. "I'm divorcing the bitch," he said, "what're you boys drinking? I just cleaned out the safety deposit box, cancelled the credit cards, and had the movers in and the whole shebang carted away. It's my pre-emptive strike. Haven't I met you before?"

By the time Sean left them, he had given Mordecai a large silver bracelet with a piece of triangular turquoise set in it in honor of Mordecai's knowing Hack Wilson's lifetime batting average. From the El Condor club they waffled along the railroad tracks into New Monterey and ended up in Chino's where they had a snack of the free tiny hot dogs on the grill there and Mordecai explained how he first found out about his melanoia. By then Mordecai had traded the bracelet for a portable TV, suitable for use in his tool shed up at Buck's house, along with a guitar. Once they'd eaten their fill of free tiny hot dogs, they moved back down the street to Beer Springs where the guitar changed into a buck knife and a pair of binoculars. Mordecai was considering a bag full of miscellaneous junk in exchange for the binoculars when he looked up and saw the fellow in the beige sports jacket having a beer back by the pool table.

"Oh," Mordecai said, "come here, Ed, I guess you might as well see this too." Mordecai walked over to the fellow's table and sat down. Ed pulled up a chair. "You're the new guy?" Mordecai said to the fellow.

"Oh yeah," the guy said, "I'm new to this bar. My name's Bret."

"This is Ed," Mordecai said, "he's studying me too." Mordecai waited as Bret picked up his beer and drank. He was left-handed. Mordecai reached over and pulled open the right hand side of Bret's jacket, exposing a pistol in a holster. Bret looked embarrassed.

"Don't do that in here," he said.

Ed Willits looked from Bret to Mordecai and back again. "What's going on?"

"Bret's doing some field work too, tonight. You want to talk to me?" Mordecai asked.

"Oh no, no," Bret said. He drank his beer and got up. "Just checking in. That's all." He nodded to Ed and left Beer Springs. He seemed nettled.

"What was that all about?" Ed said. "He a cop?"

"No," Mordecai said. "Government."

"Government?"

"Yup. CIA."

"CIA? What's he doing in a dump like this?"

"Checking up on me." Mordecai watched the pool table play for a moment. Tom Soper was about to attempt a bank shot and Mordecai always enjoyed watching Tom Soper make a bank shot. The right angle ones he did terrific. Mordecai always wondered if it was because Tom was a carpenter.

Ed was seething. "You're not going to leave it at that, are you? What is this all about, anyway. I mean, you're just a . . . " Ed didn't seem to be able to find the right word for Mordecai.

"Oh, I used to work for them, sorta, when I was in the Army," Mordecai said mildly. "I guess you might as well know this too. I was just a listener, you know. I listened to what the Red Chinese pilots said over the radio and wrote it down. That's what I was trained for. It was boring really. So when I decided I had enough of it, I went crazy a little bit so they could discharge me. Wasn't really anything much but they fell for it. I mean, nobody knows this here, so don't say anything."

"Why's this guy show up?"

"Well, they were sure I was a spy. Or a counterspy or something. See, when I went crazy there, they put me in a hospital and they went through my things. You really want to hear this?"

"Of course I want to hear this," Ed said. He was about half drunk but he snapped out of it long enough to begin *observing* Mordecai again.

"So they found this drawing. They brought it to me. They said what's this. I told them. They didn't believe me." Mordecai was feeling bored. He watched the pool playing.

"What was it?"

"It was a drawing of a harbor in Korea. I had a girlfriend there and I was studying a rare form of Japanese mathematics just for fun, it was sort of like Chess, and I was waiting for my girl and I drew a picture of the harbor above the math," Mordecai said, "you know just doodling, and they thought it was some kind of code."

"Didn't you tell them?"

"Sure I did. They didn't believe me. Then they found the sand and. . . . "

"Sand?"

"Yeah," Mordecai said, "sand from the Carmel beach. I just love that sand. I took a baggie and put some in it before I went overseas. They found it and analyzed it and reanalyzed it and tested it and then they came to me and said what is this? I told them. They didn't believe me. They thought it was part of the code."

"What did you know that was so important?" Ed asked.

"I don't know. They thought I knew something. Maybe I did. You know they have real precise tests for this sort of thing and before you become a monitor they test you so they know just how much information you can handle before you start putting together things . . . anyway, I was just a radio monitor but apparently I pulled the wrong time to go on my crazy routine or something . . . who knows? Who cares?"

"So why do they . . . ?"

"Check up on me? Oh just to see if I make my move, get my gold out of the bank in Switzerland or something, who knows. Who cares?"

Ed stopped observing and began thinking. "Christ, man, don't you see? You just reversed the tables on them. You . . . you should have been paranoid and instead you turned into this imaginary disease of yours . . . don't you see, *that's the answer!*"

"What answer?"

"Answer to your disease!" Ed was really excited.

"Why's it need an answer. It's not a question."

"Well, now we can help you make an adjustment."

"*Adjustment?* What the hell are you talking about?"

"So you can understand it. Understand how it works."

Mordecai was silent. Ed took out a small notebook and began to write down things in it. Mordecai watched him. "You know what you are?" Mordecai said.

"Me?" Ed was dividing the notebook page into four even parts and drawing arrows diagonally to each part and labeling the arrows.

"You're an understanding addict. You think that understanding is some kind of magic. You think it's real hard to understand something and you have to work at it for a long time and then the magic works. Actually that just keeps moving you farther and farther away from what you're trying to see."

Ed stopped scribbling. "I don't see what you're saying. Understanding isn't an addiction."

Mordecai looked at Ed's beer. "You understand that drink there?"

"Sure I do." Ed looked up at the beer. "It's so much alcohol and so much water and so much malt and barley and rice or whatever. I know the effect it has on my body and I know how much it costs and probably what it's color is called . . . I mean, what's to understand?"

"What's it called?"

"It's called a beer."

"What's it called before you think of its name?"

"Nothing."

"And now you understand that?" Mordecai asked. Ed nodded. "So you understand nothing," Mordecai said, "it's gotta be an addiction if you bother to do that."

From Beer Springs they went on a romp through several houses in Pacific Grove. By then Ed Willits had stopped drinking very much at all and was busy taking covert notes when no one was paying attention. Mordecai had traded the binoculars for the bag of junk and was busy exploring the junk with the friends that were in the various houses. So it wasn't until they were sitting out on a beach somewhere near Fort Ord with an old Army buddy of Mordecai's named Roy Smirle that he found out why Ed was so strangely silent during the last few hours.

Mordecai and Ed were sitting in a hollow formed by two sand dunes with Roy Smirle. He was passed out, with one arm around a gallon jug of rose wine with cocktail fruit floating in it.

"Jesus," Ed said, "was I glad to get out of *there*."

"Where?" said Mordecai. He was looking at the flare gun in the bag. He hadn't seen one of them since the Army.

"Back there in that last house in Seaside. Jesus, didn't you *see* what was going on?"

"No, what?"

"I walked into the front room and some guy had just pulled a knife and Christ, I thought we were all going to get killed!"

"Really"

"I was goddamn paralyzed. I think it was some kind of drug deal. I could have been in bad trouble."

Ed rubbed his eyes. "Goddamn, does this happen to you every night?"

"Well, every night that my melanoia is working, something like this happens. I've never noticed anything being *exactly* the same, have you" Mordecai looked at Ed but Ed wasn't aware that Mordecai was having fun with him. Ed was beginning to bore Mordecai.

"God, I thought it was a joke, but either you've got the crazy disease or you're goddamn charmed. I have never . . . what's that?"

"What's what?"

"That. There. That whump."

Mordecai was inspecting his flare gun cartridge. He stopped and listened. He didn't hear any whump. He was beginning to get very bored with Ed. Listening to him think was like listening to a record skip. "That's just Fort Ord. We're right below it, I think." Mordecai looked up at the sun rising above the brown hills in back of them. "About time for them to start."

"Well . . . I . . . " And suddenly Ed blinked at another, closer whump and a shower of sand flew over them. His mouth opened wide. "Where the fuck are we? Where'd this asshole take us anyway! Wake him up!" Ed jumped to his feet and began to kick Roy Smirle in the leg.

Whump. There was another shower of sand. Ed clambered up to the top of the dune behind them and peeked over the crest. Then he came tumbling down.

"We're right on the fucking mortar range!" Ed screamed. He seized Roy Smirle's limp left arm and began to drag him down the hollow toward the sea.

Whump! A clot of eelgrass flew over Ed like a green witch. He screamed and flattened himself on the sand. "How do we get out of here?" he moaned.

Mordecai picked up his bag and took out the flare gun and cocked it. He held it over his head and pulled the trigger. "See?" he said to the astonished Ed. Over their heads a red column of smoke was ascending and then it burst at the top and a parachute of smoke and fire began falling. "No problem," Mordecai said.

On the drive back to Monterey Ed had calmed down enough to piece together the results of his night's research. "I don't know exactly how this works or how the hell you do it, but I think you ought to come down to Stanford with me and we'll run you through some tests."

"What good would it do for me to get melanoid in a laboratory?"

"I'd like to make some more studies of you, under some more *formal* arrangements."

"I'm not sure this is a formal mental disease. I'm not sure this works under those kind of arrangements. In fact, I'm sure I wouldn't want to."

"Why not? We'll pay you."

"I don't see any point in getting melanoid for science. What will you guys do with it once you've pinned it down? Isolate some hormone or other and have everyone following each other or imaginary people or whatever once they get a shot. It would just turn into another control mechanism, Ed, and who needs any more of *those*?"

Panorama Wind Haiku

    motionless wings
a seagull faces stillness
      above the white caps

    rooted to the cliff
the cypress trees let the wind
      do their breathing

    in tidal sand
a wheel is made from two dogs
      whiffing crotches

    as I leave
a wisteria blossom
      enters the cafe

    into the fog
endless steam from the roof
      of the insurance building

    the child's hands
cannot reach the explosion
      of white wings

    the lupine seeds
are rattling in their dry pods
      the number of the wind

    I am going to see
whose breath this is on my face
      that rings the chimes

    a red-footed pigeon
looks to see what the woman
      cleared from her throat

    thunder of a jet
muffled by the whispering
      of sycamore leaves

    blackening the street
the coffin-bearing hearse
      broke up a dogfight

    look at the mime
the only movement is the wind
      through his red hair

    grassblades so tender
the litter is being lifted
      by the green, the green!

    the blue iris calls—
o wind, come blow this snail
      off my stem

continued

    diving at people
the blackbird grimly thrills
    in guarding its young

    entering my lungs
the breeze mixes with the steam
    of boiling squid

    asleep on the bus
sunlight is carried cross town
    on the old man's face

    the fuchia feels
the difference between the wind
    and the bumblebee

    those eastern clouds
reveal a high mountain range
    before rock was born

    jarring the window
the wind tries to free the air
    locked in the glass

    to get down the wall
all the cockroach has to do
    is merely let go

    a cramp in his calf
the tourist hammers his fist
    in the dawn hotel

    all night garbage
clatters down the metal chute
    no bottom sound

    the moon cannot see
what the sound of the foghorn
    feeds to the dew

—Fruud Smith

Brent Richardson

CROSSING THE TRANS-TEHACHAPI HIGHWAY
Peter Coyote

Crossing the Tehachapi Mountains, which gird the loins of Southern California like a reversed chastity belt and lock the creeping fingers of Los Angeles' northern sprawl out of the Central Valley, requires an imaginative leap. The highway I refer to is the San Francisco to Burbank air corridor I travel weekly seeking work as a migrant laborer in the film industry. Traversing it regularly and rapidly leads one to the conclusion that, rather than connecting disparate sections of a single state, it separates radically different countries.

Anyone who's seen TV or foreign films knows that urban settings, industrial slums, automotive congestion, grandiose hotels, shoppers and sightseers are common everywhere. So, one does not expect the differences to be apparent in these areas. It's the collision of geography, weather, plant and animal life, vision, and language that exert inexorable pressure here. These transform the human inhabitants subtly and eventually permanently. Southern Californians are not who we are up north.

I leave San Francisco dressed against its aggressive early morning chill. Fog steams off the bay, caresses San Bruno mountain, moistening the feathers of gulls, terns, and sandpipers pecking and bobbing their way along the broken rocks on the Bay shore.

The early morning freeway feels relaxed and expansive. I park my truck in an off-lot and deliver myself to a naugahyde-paneled van which will ferry me to the airport. It is regularly driven by a Fiji Islander. If life were symphonic, this plastic-Pacific mix would be a thematic pre-figuring of the way Los Angeles feels.

The airport itself is truly nowhere. Tons of landfill and concrete have nullified consciousness of the surrounding marsh and Bay. Ghosts of cordgrass and shrimp, mussels and millions of shorebirds haunt the distracted travelers humping luggage through the neon-lit, tiled corridors. All physical, organic presence besides humans (the parasites they carry), has been banished. These corridors are the domain of man's imagination—or rather that part of it which competes with Creation—and, in this tidy playground at least, it reigns supreme. "White courtesy phone please. . . ."

Inoculated by this environment, and scrubbed clean of all natural frames of reference, we board the plane, attend the mindless ritual of oxygen mask and exit drill and are slung aloft so that the earth can turn under us.

From the air, Northern California is a spectrum of rich greens. The houses and freeways are effectively limited to narrow corridors by the geological realities of the Bay and Coast ridges. Heading south, the plane banks gently east, following the Coast. A few tremors of turbulence and we edge over the last parallel ridge into the Central Valley.

The colors change to dun, sienna, ochre, rust, beige, grey, and tan. To the east, the Sierra-Nevadas are usually masked with clouds. Directly below, dirt roads meander everywhere and nowhere in particular. They don't connect towns, dwellings, or recognizable locations. They cross hills, branch about the flat plains, strike odd angles and describe perverse tangents. Occasionally they dead-end at a pond or water tank. The pilot usually calls attention to Fresno as we pass.

Fifty minutes south of San Francisco, the plane jerks and shudders, the seat-belt warning light goes on as the air currents—like sentries—rush off the Tehachapis shaking and frisking down the plane as it crosses their border. The great transverse range rolls, humps, and chasms below, dividing the open fields from the endless mega-city stretching south. Borders are protected here. Value systems change, language changes, and *you* better change if you're going to survive. Not even the plane is immune.

The pilots must be Los Angelenos because they always make a joyful, stomach seizing roll over the hills before dropping the plane like a couch onto the Burbank runway.

At eight in the morning it is already hot. The air is a sultry beige carpet. Trees are solitary events called Palms, designed by Dr. Seuss. The only people hurrying are immigrants in polyester business suits, carrying attache cases and garment bags. The natives appear to be standing still in sun-glasses, looking and waiting for something to happen.

The San Fernando Valley surrounding the airport is a three million soul ante-chamber to Los Angeles itself. Another canyon-furrowed ridge, probably a western spur of the San Gabriels, separates this spot from Los Angeles and its unbroken roll into Mexico.

San Francisco's architecture and a good piece of its ambience is Yankee.

Its chilly clime and gritty fog, and the fact that it is a short downhill roll from the gold country, induced eastern immigrants to stay put there. Indigenous people were too smart to live in its windy damp, so San Francisco was a tabula rasa for the fantasies of newcomers whose standards of elegance were New York, Boston and Philadelphia. They erected Victorian monuments to Mammon that looked exactly like the big homes they had envied or occupied in Brattleboro, Back Bay, Providence and Fall River. The city was built so quickly that even today, the pervasive and eventually dominant influence of Pacific migrants and cultures is played on the sets of an eastern theatre piece.

Los Angeles is different. Ethan Frome wouldn't have known what the hell to do with a maguey plant. The light has no edges and is pitiless to color, reducing them all to pastels.

My friend, Redwood, came down from the Trinity Mountains up north and married a Japanese wife here. She teaches in a day care center where 28 languages are spoken. I have yet to hear English on the sidewalk outside their house. Their building is behind a Vietnamese Buddhist temple and the sound of gongs and chanting and the heavy scents of incense wafting through the window make it impossible to identify one's location off-handedly.

The air is soft and people expose more of their bodies to it. Nubile, Pan-Pacific beauties, whose innate grace and sensuality annihilates centerfold sexiness with a casual backhand, shuffle along in zoris, cut-off jeans and fluorescent halter-tops. Filipino fashions, short sleeved shirts worn outside loose trousers, work for the men at Ninth and Normandy who smoke and talk casually amidst the Thai and Korean restaurants, cut-rate linoleum and liquor stores.

Spanish is the lingua franca. It's not unusual to meet Asians who speak Spanish as a second language and no English. Some Japanese tourists who had studied English for their trip abroad were disconcerted recently to discover that many Los Angelenos could not speak the official language of the nation. In San Francisco, if an Anglo speaks Spanish, even badly, Latinos and Chicanos are surprised and usually try to help. Down here no one waits. This may be the Tower of Babel, but in L.A. the foremen speak Spanish.

Beverly Hills and Brentwood are harder to describe. Imagine a Spanish Mission rammed by a semi-truck full of Italian modern furniture. There is a sense of space around these posh homes that the mansions on Broadway and Pacific don't have; a roomier, more relaxed kind of feeling in the expansive lawns, old trees, many flowers and cool, scented shade of ramadas. Greece, Italy, Portugal or Spain perhaps, but not Scotland. Definitely not.

For all the impress that weather and bio-geography exert subliminally, L.A. people pay almost no conscious attention to it; not even to carcinogenic air. It's chilling to be driving down the pike and hear the radio announcer tell you in the same tones he reserves for the time and temperature, that buoyant, best-of-all possible, "just the facts folks" chirping, that "the air quality is hazardous today and you should refrain from unnecessary activity"—this while you're driving to work.

It's no wonder that dislocation and surrealism are at home here. The city does not draw its life support from anything indigenous except the oxygen. Water comes from the Owens Valley, Mono Lake, Yosemite and the Colorado; energy from as far as the Four Corners. With natural water sufficient for about one-tenth the present poulation, the rest is supplied by engineering miracles, economic and political piracy.

Illogical, cancerous, swollen and bloated as it may be, there is something awesome about the scale of North America's largest city. That must be said. Pulling onto the San Diego Freeway at seven a.m. is a staggering experience. The on-ramps have stoplights metering the flow of traffic. When yours turns green you merge with a conveyor belt of roaring, rumbling, sputtering, and screaming automobiles, packed solid as sardines and traveling at 65 miles an hour. As far behind you as before you, as far as you can see, a procession of machines holds dominion. Individual significance shrinks to naught; you become an integer of ceaseless activity, a member of a species with an energy that is larger than petty, facile judgments. To see this you'd have to be prepared to perceive the beauty in an atomic explosion. It's not easy or instinctive to do, but it's there, and it's here too.

L.A.'s current course is suicidal we know. The economics that make it feasible for me to travel 800 miles a week in search of work; that forestalls people paying the true cost of their water and life support; that substitutes mechanical reproduction for human skill, is sure to exhaust itself. The fundamental energy *behind* it, however, is inexhaustible. The billions of intentions, greeds, aspirations, desires, and dreams; the billions of horsepower and burritos, Buddhists and Bahais, Chicanos, Latinos, Chinese, Vietnamese, Anglos and Blacks, Hawaiians, Samoans and Thais et al. are an extrusion of the universe itself. It could as easily manifest that energy here constantly and reverently as it now does raggedly and rudely.

As a man who "acts out" human possibilities for a living, I have profound respect and some understanding of the fundamental mutability of human beings. I can apprehend L.A.'s vast energy shifting its present course and driving different kinds of endeavors and understanding. Ignorant and lethal as it may be it is also the cauldron from which wisdom and compassion might manifest itself in appropriate forms for the time. I love it because of this and because it is large enough to be an appropriate symbol for Life. As such, ideas about the God of Life must include the God of Death. That God, in all His irridescent, fluorescent, and smoky plumage is awesome and beautiful too and wisdom can as easily issue from his gnashing, clacking teeth as it can from the mouth of his pacific twin.

The final flowering of Europe, Democracy, Capitalism; the karma of the country's founding in slavery and imperialism is right here in L.A. All the contradictions and follies and possibilities of this current world imagining are chugging, slurping, sucking, zapping, rotting and pumping away without direction or organizing myth. It's the Fool's deal now, but it could shift to Christ, to Mammon, Buddha, Pantheism, or annihilation. It's pure agar, ripe for inoculation with a compelling vision. It's the present age. It is as far from the climate and style of San Francisco as it is from the emergent regional consciousness that blooms by the Bay. These are the two ends of my Trans-Tehachapi highway, and the two realms of my migratory life. They are also the two pans of a scale marked Past and Future. It may soon be possible to abandon the road altogether and settle down to a livelihood and vision with its feet wet in the waters of home.

July 1981

Los Angeles: Despues de Treinta Años

My father
   walked these streets
Staying in cheap hotels
   looking for work
   fifty years ago
And he and I both
   driving trucks
All around these
   alleys
Delivering milk and
   cottage cheese
   butter
     ice cream
To restaurants forgotten now
Starting out at three a.m., a
   little boy at his side
   sitting high in the cab
   of a big truck
One day I got my own
   so I know,
     I know these streets.

Calles de los angeles
   thirty años
   han pasado
Un niño yo
   in streetcars
All over this damn
   gran ciudád
No, no ciudád
   pero sprawl
   la orígen del
   concepto
   "sprawl"
All over this place
   in tranvías
   I used to go
Looking for
   souvenirs.

Gran cambio en
treinta años
But todavia
the same
Mas Mexicanos
ya no tranvias
only buses
Breathing diesel
in L.A.
diesel en vez
del aire

Mas Chinos y Asianos
mucho mas
Chinatown sure has
grown
where is Little Italy?
Bus loaded con niños
Mexicanos y
Chinos going to
school
The Asians are invading
los barrios abandonados
de la raza.

L.A. object of hate
yes I thought
I hated her
Pero ya, no sé por
cierto
Memorías flood my
consciousness
So many times
I've been at this
very sitio
Un niño volante
exploring
bookstores y
watching
Los ancianos, winos
y good looking
black women
From the south side
A meeting ground
donde no one
meets,
sin conjunto

Stains on dirty sidewalks
like jackhammers
Indians bashing brains
On cement walls
their own matter
splattering
Blood marks forming
Hollywood Boulevard
halls of fame
For unknown stars
guided tours
Gray-Line
Telling which celebrity
puked there
which one
peed in that corner
Over there
one had the guts to escape
On the front of a car,
salsipuedes.

America will die
when Indians die
Are we still alive?
or is it all a lie
Rotten corpses
refusing to admit
In front of Temple Baptist Church
that it is all gone?

Sixth and Spring
oeste to Plaza Pershing
Integrated city
older, broken whites
international winos
Third World youth
You can almost tell
a person's race
by the expression
on his/her face.
Why do the young whites flee
leaving sallow winos
rag pickers
abandoned ancianos
only the poor
To be visitors
in a Mexican/
Chinese/Korean/

African/Indian/
Filipino/Mestizo city
donde todo el mundo hides under
rags por noche
Cuando cucarachas are
looking for crumbs.

On the streets of
downtown L.A.
Barnacled mossy hull of a boat
beset by waves
Rocking between junkies
pensioners, drunks,
children in arms of strong-faced women
out-of-work men looking
for something lost
An Indian eagerly embracing una medium-brown mulata
en la plaza
across from the Biltmore
Race not too important when a city lawn
is your living room
and your hand wants
to feel something soft.

Crowded life-boat
nine-tenths of the people poor
half or more very, very poor
three-fourths non-white
The whites are all at work inside
rushing out at five
to jump into
Speed-boats
escape boats
that leave the city
empty of suits and ties
by six.

Chinatown has really grown!
Brown school kids rushing out
to stand in aisles
of crowded buses
The barrio has expanded al norte
Chinese crowding in behind
Mexicanitos y Chinos bright-faced
looking good
junior-high and younger
not broken yet
Could be la futura de la ciudád
if the city's death
can be delayed
A while longer
with the arc of poverty
catching more and more
If you haven't got the cash to fly to Mexico
no matter
Mexico has come
to meet you,
And Asia too.

And from the southside
twenty-miles of ghettos
Come the Blacks
los Mexicanos tienen el este y norte
los Negros el sur y suroeste
los Coreanos, Centroamericanos, Filipinos y
Japoneses el oeste
los Chinos and Vietnamese squeeze in where
they can
same con los Indians
The better class of Indios are way out
con los okies
y hillbillies
Los pobres
crawl in any hole
they can.

Thirty years
Makes me edgy
     been away so long
Lot more poor people now
    War on Poverty?
In L.A. that war
    is a rout
  Battlefield shifting
    block by block
    Spring Street: lost
    Broadway: lost
    Hill Street: lost
    block by block
    all lost, the whole downtown
Public monuments and vast parking lots
  where no one lives
    is all that's left,
      only from eight to five.

Howard International Hotel
  conveniently located at Skid Row
   twenty dollars a week
  ambulances instead of taxis
No airport limousines
  just police vans
  old people
    families
Inside behind locked doors
  watching "The Jeffersons"
   on blank walls.

Treinta años
  good thing puedo leer en español
   used to wander all over these streets
Y ahora?
  lo mismo and yet so . . .
   depressing
To see these white-man cities
  abandoned to
   los morenos
Thinned by agribusiness and multinationals
  and computers y maquinas
   from homelands
Driven like sheep to cement corrals
  to crowd and multiply
   some hope for miracles
Others stare straight down.

Sturdy campesinos
how long does L.A. take?
How long to
break you in two
How long to
take your sons
How long to
take your daughters?
La cultura del pueblo
dies slowly in the grip
of gangs y drogos
The culture of the streets
sometimes outclasses
even Confucius.

Oceans and oceans of houses
ghettos y barrios
in cada dirección
ten miles here
thirty miles there
The whites are mostly out at the edges
hanging on mile by mile
and then retreating con su dinero
to San Juan Capistrano
and the Desert
Gasoline prices on their minds
freeways their
lifelines, but
The strands are breaking.

Water from northern Califas
the juice of white urban flight
Without more water
they would be stuck in L.A.
Water for swimming pools and
artificial lakes
Making the desert into bedroom suburbs
with water
To escape from
wetbacks and niggers
Robbing the taxpayers
draining rivers
Drying up the north
for sprawling flight
Building highways to heaven and a
gas-lit barbecue
but the price of gasoline gets
higher
and
higher.

L.A.
    ciudád sin alma
Real estate promoter's
    spawn
Paradise fouled
    ciudád en dos mundos
From the windows of the Biltmore
    rich decorations at my side
I can see
    multi-colored flowers
    of Pershing Square
Trampled down
    roots torn loose
Laying on the grass
    pensioners arguing heatedly
With indifferent winos
    some don't even hear.

Downtown L.A.
    off-limits now
Freeways encircle it so that
    middle-classes
Are spared its sights and sounds
    buses cut through it
    that only
The poor and non-whites ride
    Go around!
    Go around!
Whispers heard in suburban malls say
    "it's only a bad dream"
    Go around!
    Go around!

A city surrendered to the forgotten
    a city forgotten.
    A city I wish I could
    forget.

—*Jack Forbes*

WORK

Worker,
  Put your shoulder
To the sky
  —William Garrett

HARVESTING THE TRASH
Bob Beatty, talking (Managing partner, Berkeley landfill)

I first got into trash when I was 17 in 1965. I was working as a timekeeper for a general contractor, John S. McQuade, in Philadelphia. Mr. McQuade's company was under contract to remodel one of Philly's oldest department stores into a shopping mall. Well, Cleveland Wrecking had a subcontract to trash the top eight stories of the building. For the next six months, seven days and nights a week, Cleveland took truckloads of material from that grand old building to the dumps. They worked three shifts, sometimes a hundred men to a shift. I kept their time and made out their checks each week. When they started taking down the main courtyard, there were these pink marble tiles that were being hauled away. After work, I'd go salvage those tiles. I got about 5,000 of them and sold the lot for a dollar each. That gave me a feeling that things do not end at Sears Roebucks, that trash is valuable.

I also got into salvaging a little bit as a trader over in Vietnam. I was in the company supply depot towards the end of my tour. My job was to keep the company well supplied with weapons, ammunition, and also clothing and food. To be really good at that you have to be a bit of a wheeler dealer. You have to be able to trade. That's what being a scavenger is all about. It's trading. You have to be able to know a good deal when you see it, pick it up, and trade it with somebody.

I came out here in 1975 and started working with a guy taking down piers in South San Francisco. He had the good fortune to buy a redwood drydock from Bethlemen Steel for $10. This drydock was 600 feet long and was used during the Second World War to repair submarines. Bethlehem Steel saw that it was no longer of any use to them so they wanted to get rid of it. They sold it to this eccentric German man, Claus von Wendel, for ten bucks. He was an amazing character. An artist of sorts. I was the marketeer for the group. My job was to sell the salvaged lumber.

From there, another mate and I from the pier operation started Ohmega Salvage. We began taking down buildings at Treasure Island, salvaging the lumber. After doing the Ohmega trip, I started Flight Salvage. It was more of a philosophical adventure because it employed ex-cons, trying to salvage their lives. We were salvaging people with salvage materials. That was real interesting. We took down a department store in Stockton and at one time had about 50 people working on the job. I was getting calls from San Quentin and Folsom prisons saying that they had potential parolees. All they needed was a job. So I said sure, "Bring me your homeless, bring me your needy." Unfortunately, the philosophical adventure was a financial disaster for myself and my other partners at Flight Salvage. It turned out we were hiring more people than we could afford.

I began using the Berkeley dump to get rid of our left-over wood. We found out that the dump was coming up for renewal of the contract. So, we went out and pressured and lobbied the city to put it out on the bidding block. We teamed up with a local demolition contractor, Bay Cities, whom we had previously done some salvage with. We had taken down this building at the corner of University and Shattuck with them. It used to be the Owl shoe store. So together we put a bid on the Berkeley dump and got it. We took it over August 1st, 1979.

Running the Berkeley landfill for Bay Cities has been real interesting. It's like running a farm. Only instead of crops we harvest trash. We pick the trash. Bury it. Move it around. Dispose of it. It's interesting to watch people when they throw things away. They almost have a guilty look on their faces. Yesterday a guy came into the dump in a station wagon loaded with trash. I was standing to the side of his car and he was throwing these things out. He had a hand lawn mower and I thought maybe he wasn't throwing it away. So I asked him, "Are you throwing that away?" And he said, "Yes I am." I said, "What's wrong with it?" He said, "The wheel makes noise, it creaks." I said, "Oh." I didn't want to say anything negative because he might say, "Well, I don't like it. I feel bad enough as it is. Don't make me feel worse."

Then there are the people for whom coming into the dump is like coming into a no-man's land. They think there's no law out here and everybody is into themselves and you might as well just pack a pistol and not worry about it. People come into here racing 60 miles per hour. They wouldn't do that going down their street, but they think that there's no chance of any policeman catching them out here. Attitudes at the dump cross all stratas of social, cultural, ethnic, and racial lines down here. The most ignorant are the educated, cultural ones. The people who act with a certain amount of awareness are the people that have been scavengers, who have come from that end of the spectrum their whole life. We have, for example, a lot of old black guys who come down here. They are always real nice, courteous, friendly. But then there are the other people who come up in their little Volvo, Mercedes, or their Chevrolet Capris and they just can't understand why they have to pay to dump their trash. Then, as soon as they pay, they want to go around the back of the office and unload everything. Those people, that I'm sure are college educated, say "Oh, I just dump anywhere?" I say, "Dump anywhere? This is my house. I'm trying to keep a clean house and dumping anywhere is like dumping cigarette ashes on the sofa or spilling the beer on the kitchen table." I think to myself, like Garfield, "You slob, I'll dump my trash in front of your house."

People also throw away bad memories at the dump. It's always personal, personal, personal. Lovers breaking up and they want to get rid of some furniture, photo albums of their ex-wife or husband or whatever. Or somebody just moves out of a house and leaves everything. People also clean out houses of their parents or grandparents that just died. They throw everything out, totally useful items. But they're just cleaning, taking care of the estate. Yesterday, one of the fellas got a diary that was kept by a kid who was in grammar school in 1910. And this diary was meticulous. It was a school bond type of copybook. He did a journal of Haley's Comet. It was the last time that Haley's comet came in. He had real nice drawings. So if you want to know astronomy come down to the dump. It's a galaxy, a constellation of things.

Another thing is that we're retrieving metals from old electric irons, toasters, and frying pans. The glut-end result of consumerism, of all this manufactured garbage that we're getting sold. The dump is the depository for planned obsolescence, the non-durability of products. We get stoves, refrigerators . . . and once in a while, a car.

The city gives lip service to recycling. But it's only lip service and that's standard throughout this country. It's different in Europe and Japan because they don't have that much land to begin with. So they don't think of landfills as a permanent solution to their trash problem. Here, it's just the opposite. We have so much land we think of landfilling as a quick solution to a big problem. But that's changing. Urban land is scarcer. Pollution. The whole thing of hazardous waste. Love Canal was a landfill. In fact, I was reading in *Not Man Apart* where in Kansas a society of civil engineers are pressuring the state government to make it illegal for landfills to accept hazardous waste. Meaning that right now landfills do accept hazardous waste in Kansas. California has stiffer regulations than the federal regulations concerning hazardous waste.

I was thinking about putting an ad in the Wall Street Journal and having a seminar at the dump where executives could spend a day or two. I would have them come in their overalls, or their tennies, whatever they feel better in, and take in the view; the wind, the dust, and what people actually throw away. Then, at the end, give them a copy of the Global 2000 Report and have a few people talk to them. Talk about putting somebody close to their environment, that's where it's at, right here at the dump. What you see at the dump *is* America. Items from I Magnin, Macy's, Ace Hardware, REI, Taiwan, Bolivia, Chicago, the whole planet. It's a future archeological site with all the artifacts of our culture.

Maybe we should have an area at the dump that would be zoned just for sculptures, like the Emeryville mudflats. I was thinking of renting a gas station across the street from Fat Albert's for about a month and showing the public what things are thrown away. We could give interpretations of what these discarded things mean. For example, today a Chevrolet sign came in from a dealership on San Pablo Avenue that closed down. In twenty years they're not going to be making Chevrolets anymore, at least the way they make them now. Also, things like medical supplies. Kaiser Hospital once a month throws out medical supplies, medical drugs that have passed their expiration date. They throw them out because they may have only 80% potency. Frustrated with the waste I recently called up Casa El Salvador. They came and gratefully took some of the medical supplies. The last I heard they were on their way to New Orleans. I'd like to see somebody say something about that, where it's too hot to handle. Sending "expired" drugs to foreign countries is big business. Is it better that those supplies go into the dump and be buried instead of helping somebody? I think not.

Berkeley deals with recycling in a better way than most American cities. But even it falls short of the need for a better solution. The city government basically gives lip service to the Ecology Center, the Community Conservation Center, and the actions of Urban One at the Berkeley landfill. When the landfill closes in about three years the city plans to build a transfer station and incinerator. Berkeley spent half a million dollars on just the plan, but people are starting to see its shortcomings. The city is planning on taking all the trash that can't be burned and hauling it out to Livermore or Richmond in trucks. The people at the gate will be paying $20-30 a load for that hauling. Right now, we get some of the trash from Oakland because the people in Oakland don't want to pay the rate of Oakland scavengers to haul stuff to Livermore. We should be recycling more now, so that Berkeley can keep its landfill open longer. Then we won't be dependent upon the prices of outside forces, namely Oakland Scavengers and Richmond Sanitary or some new dump in Livermore.

With a few minor changes we would be able to adapt and recycle more. We already have been slowly adapting to that. When we first took the dump over, everybody was dumping pretty much in the same area. Right now, we have the city dumping material in one area. That area would be an ideal place to possibly recover methane. We also have a composting area for brush and lawn clippings and an area for breaking down ferrous and non-ferrous metals.

If the city allowed it, we'd like to have another area for wood waste. There are firms who want to have us chip it up for fuel. They'll pay us $14 a ton, right now. In a year's time it may go up five dollars a ton. Instead, Berkeley will three years down the line pay somebody else to haul it out. That's crazy.

The city isn't interested in genuine recycling because it's innovative. There's no precedent. Bay Cities could do it if they would put out a capital investment. But they are very conservative. They want to be able to guarantee their capital return. To my way of thinking, there's a certain amount of risk in any business. But it's just too innovative for Bay Cities, too. They would just as soon bury everything. So now the city is talking about this $30 million Transfer Station when the dump does close down.

One viable alternative would be to keep the present landfill open. Make it a resource recovery depot where we would only landfill items that could not be recycled. It's now scheduled to be a park. But, there's no charter or anything to make it a park. I don't think we need a park in that area, specifically, because it's very windy. If it becomes a park I don't think it will be used to an optimum. As far as boating goes, they would have to built another marina. In terms of fishing, we get some fishermen right now. But, on an average day, there's a dozen at most. I don't think making the place a park would increase fishing that much. The landfill's highest use really would be to continue to use it as a place for processing trash, composting, and salvaging. We could run it as a trash farm where the yield would be recovered materials. Instead they are planning to build this monolithic burn-combustion center and Transfer Station. What if ten years from now we're not producing as much garbage as we're producing now? That center will be obsolete. It won't be able to burn *anything* without a large volume of combustible trash. Even more, at full capacity it will still have to be fired by fossil fuel. It won't be able to burn entirely on trash. So we are going to have to be somewhat dependent on outside energy to keep it going. And the cost of that right now is about 30 million dollars.

We could instead have a trash farm for a total investment of less than $1,000,000. The thing to keep in mind is that we'll have a product with that investment that we'll be able to sell. For the metals, right now, it would be a little bit harder because scrap metal prices are really down. Still America produces far more scrap metal each year than it consumes. The biggest customers for scrap metal are Japan and China. It comes back in the form of Hondas and Toyotas. What we should do is set up a clean smelter at the dump and make some electric cars or some trash cans like they do in Japan. In Japan they have a smelter right at the dump. They make nails. I would really look into the possibility, as far as the scrap metal, of coming up with a viable product that we could turn around and sell to the public. The same for composting and wood chips. Also, household goods: all the chairs, beds, mattresses that come in. For an investment of less than $1,000,000, we could have a product. Something we could sell, be able to get an income from instead of building a gigantic Transfer Station for $30 million. For $30 million the best the city hopes to get is maybe some steam for electricity. Even that is nebulous. Worst of all, with the Transfer Station we will be paying somebody to haul that trash somewhere else. The whole thing is negative. It's just moving the problem somewhere else. We could instead stop the problem, come up with some kind of creative solution.

Private investors are interested in the idea. But they're looking for a commitment from the city and also from Bay Cities. But the people of Bay Cities unfortunately are not missionaries in that direction. They're basically highway people. They're into paving and grading. One of the large broker houses, Paine-Webber, Inc., is interested in doing something with this, but they want a contract. They want to see where our relationship is with the city. But the Public Works Dept., which controls trash decisions, isn't interested. They have consultants who "know better." They're professionals in solid waste management, but in only one area. They believe in burning everything first. Anything left over will then be considered a renewable resource. There's also some other professionals that are going to be bidding on the Transfer Station. Their proposal will be one of minimal burning and maximizing composting and resource recovery. But that's still not the best solution.

The bottom line for the city in its decision presumably will be who gives them the most money. Tied into that is the guarantee that they will have somebody who will be able to take care of the trash problem. Because the city still looks at solid waste as a problem, as a health hazard. They think that it's just going to be creating itself perpetually. I think, with the reduction in packaging, and conservation that is happening, that garbage will be reduced. The people in the Public Works Department don't think that way. To them I say—rubbish! I think the dump should be under a Department of Resources, not Department of Public Works. Maybe there could be a liaison between that particular agency and Public Works. We have to think of trash as a resource instead of a problem.

Dennis Hayes, from the Solar Energy Research Institute in Denver, Colorado, has shown that right now taxes on the shipment of scrap metals and recyclable material, paper and cardboard, are higher than the taxes on the shipment of virgin material. Also the same on depletion allowances where the company developing new resources gets more of a trade-off than one recycling something. That's a counter-productive policy if you're interested in genuine conservation and an efficient use of resources. It's also a proven fact that it takes less energy to recycle scrap metal or salvage material than it does to produce from virgin material. But there's such a strong lobby from the virgin producers that recycling and conservation are largely ignored by the government. We all need to realize the party's over, that the planned obsolescence attitude is, itself, obsolete. Even though business is bad right now, industry keeps saying there's going to be a recovery, that things will be back just like they were in the '50s. Well, that's an illusion. That's not going to be happening. There may be a temporary relief for some people, but as far as long-range, the party's over. We have to shape up and start to really harvest our trash.

Offices

Morning.
            People march
To music too unheard &
Vanish just as fast.

           *

Muzak—
   the muse is sick,
music / worse than war, worse
    than peace, the sirens are
the true, secret & ultimate
    source & mystery
of poetry, said the
    poet to the actor in
the back of the café.

           *

God smiles at me when I work, but He
loves me when I sing, sure, so,
la, ti, do, but if I hear the Muzak
play *Manha de Carnaval* one
more time, I think I'm going to scream.

The new secretary on the farthest desk, a poet, smiles
on everybody & everything that passes by her like
a flower in the Spring after hard rains.
A few months later, her smile's not gone, but replaced.
This one's a deeper, darker line from her nostrils to her chin.

*

Winter to Spring: Mooney exchanges the felt
hat for the plaid. A fifty-year-man,
he can speak about anything
in terms of departments, benefits,
officers & the Board, sports,
or the weather, never saying me:
a perfect model of the self-referential
self-effacement of bureaucracy.

*

During the 1980 Soviet May Day Armed Forces Parade, a small jeep
with four civilians was observed following the massive display
of missiles, tanks, & troops. When asked who these men were,
Brezhnev said: Those are economists;
you couldn't believe the destruction they can cause.

*

A wierd laugh carries down the hallway like
a hyena in a zoo.
I wonder what they sound like
when they cry.

—*Gary Gach*

HOME ON THE SONOMA RANGE
Leonard Charles

Five years ago, I didn't believe that coyotes killed sheep, or rather, didn't believe they killed more than the odd stray or weakened animal. A self-avowed ecologist, I endorsed the environmentalist belief that coyotes preferred native prey, while sheep kills were the infrequent acts of "bad" coyotes (usually an old or sick one, or one who had lost a paw in some previous encounter with a trap).

Agreeing with John Muir, I believed sheep to be "hooved locusts," the products of sheepranchers who themselves were notorious for the wanton destruction of wildlife, and overgrazing much of the American West. On the other hand, coyotes were noble predators, wild—even magical—beasts, and part of the balance of nature. I will always remember sitting on a Sierra peak in the full moonlight listening to coyotes explode a neighboring ridge with their searching screams while the short hairs along my spine bristled. In any debate over the relative merits of coyote and sheep, I knew whose side I was on.

At that time, I lived with a group of people on, of all places, a sheep ranch in the rough coastal hills of northern California. We operated a small homestead amidst thousands of sheep belonging to a neighbor who held the grazing lease on the ranch. Though he was considered one of the foremost sheepranchers of the region, his approach to land and wildlife management did little to alter my feelings regarding sheep or sheepranchers. The ranch was overgrazed (evidence by the spread of thistles, tarweed, and other wasteland weeds, gully erosion, and landsliding); brush and trees were bulldozed from lush hillside flats and pockets; and "pests" (feral pigs) and predators were routinely killed.

When this neighbor retired, we decided to take over the grazing lease on the ranch in order to control how the land was used. The ranch was under an Agricultural Preserve contract which required the owner or lessee to produce a specified agricultural income off it each year. Sheep were the most feasible means of producing this income, so we became sheepranchers. While none of us had a strong desire to raise sheep, we decided that if we were going to do it, we would do it right—commercially raise sheep in a manner that enhanced the health of our land.

Our first year was quite successful despite many errors and oversights. We produced a lamb crop of 90% (90 lanbs for every 100 breeding ewes) in an area where anything above 70% is considered good. During our second year, coyotes moved into the area. That second year our lamb crop dropped to 50%, plus we lost 10-15% more ewes than the previous year. This is our third year, and the lambing season is now in progress. So far, we have lost 10-20% of our ewes plus an as yet undetermined number of lambs. It is always difficult in the rough terrain of this area to determine the cause of death of an animal, because the birds and other carrion eaters make short work of any carcass left on the range. If a complete carcass is found, the determining marks of a coyote kill are puncture marks in the skin of the throat, as the coyote typically kills the ewe by grabbing her throat, crushing the thorax, and causing her to suffocate (it is difficult to determine the cause of death on small lambs). Of the seven carcasses I have found this year where some determination of death could be made, all had been killed by coyotes.

I found a dead pregnant ewe from which the coyote had eaten a mere pound or two of the choice internal organs. Above the big spring, I discovered three freshly killed week-old lambs. A few days later I saw a ewe, the wool on her throat stained wtih blood, wandering the hillside in shock. These animals were our responsibility and we had expended much time, energy, and care on them. Impotently watching them die made me regard coyotes differently.

"What to do about coyotes" became a personal problem rather than an "issue" (actually it is a problem for all of the people with whom I live, but the opinions here are mine, as some feel differently about both sheep and coyotes). Sensing that the typical rancher's solution of killing the coyotes at any expense and the environmentalist contention that coyotes are not a serious problem were equally in error, I was left in limbo. One moment I felt like taking my gun and scouring the hills for a coyote, the next I felt it was the coyote who belonged here and not us and our sheep. One night after I returned from finding yet another dead ewe, I sat down to think about coyotes, and I realized that all I knew about them were generalizations and simplifications produced by the arguments of specific interest groups, the validity of whose arguments could not be trusted. The solution, if there was one, seemed to lie in finding out as much as possible about coyotes in general, and specifically in this time and place. It was necessary to move beyond the gut feelings that swung me to and fro like a pendulum and do some thinking, some research; this correspondence relates some of that exercise.

The record on the rancher's side is totally indefensible—two million coyotes killed by government agents between 1915 (when the Federal government became actively involved in "predator control") and 1946. After 1946 the slaughter got worse due to the introduction of 1080, the new "superpoison" so potent that one ounce can kill 20,000 coyotes. Theoretically, the poison was strictly controlled regarding who could use it, how it could be used, and how much could be put out in a given area. The controls were meant to insure that the primary target of 1080 would be coyotes and not "innocent" species. However, the abuses were so gross and widespread that in 1972 President Nixon, of all people, banned its use by a Presidential Order. Later that year, the EPA withdrew registration for all predacides (poisons to kill predators), effectively ending the "poisoning of the West."

The results of that era of poisoning (an era that extends back to the mid 1800s when strychnine and other poisons were used) can never be calculated, but we do know that coyotes have been extirpated from broad stretches of their ancestral range (the high plains) and that many other species of predators, and even non-predatory mammals, have become extremely rare throughout the high plains and many other parts of the West where poison was used. The record is hardly surprising when one considers that throughout the nineteenth century strychnine baits were sown like seed grain across the plains. Consider the 25 pounds of 1080 illegally acquired by a Livestock Association in one county in Colorado, recalling that one ounce can kill 20,000 coyotes. One Federal official, when apprised of this acquisition, gasped in horror that 25 pounds was more than the government used for a whole year in the whole country. Such horror stories are legion, and it is not surprising that the mink, fisher, marten, and ferret populations have been destroyed (ferrets to the point of extinction), nor is it surprising that one rarely sees a fox, bear, bobcat, or mountain lion in areas where the poison was used.

The ban on predacides by no means stopped the U.S. Fish and Wildlife Service (through its subsection called the Animal Damage Control—ADC) from killing coyotes. Using traps, aerial hunting, dogs, calling, and other means, the ADC killed 605,000 coyotes between 1970 and 1977 (this is the number reported by the ADC for its activities and does not include coyotes killed by other agencies, ranchers, varmint hunters, or fur trappers). So in 1976, while the ADC killed 99,000 coyotes, 173,000 were killed for the fur trade (a coyote pelt goes for between $30-50) and an unknown number were taken by private hunters, never found, or not reported.

For the ranching industry, which would like to see the coyote on the endangered species list, these deaths are necessary to protect their "marginal" industry. To listen to ranchers, one would come to the conclusion that coyotes were entirely to blame for the many problems of the sheep industry. The coyote is the fall guy, not the economic system that forces ranchers to produce large numbers of animals for a marginal profit. That economic system is never challenged (in fact, ranchers are usually counted among its staunchest proponents) because "there is nothing that can be done about it." You can do something about coyotes: there a person can act, can make a difference.

The ranchers point at the spread of coyotes across the continent as evidence that they are doing no great harm to the coyote population. Coyotes are now found from Costa Rica to Alaska (where they followed the overland-bound gold miners, feeding on their dead mules). They are now found in every state in the country except Delaware. The ranchers contend this expansion of range has too often come at the expense of their sheep, and that coyotes do not "belong" in much of their present range. As our local trapper puts it, "The Sierra Club is a great advocate of the balance of nature, but for the coyotes in this country, I'm the only balancer of nature."

To further bolster their claims, the ranchers point to the drastic decline in the number of sheep in this country—28,849,000 in 1960 and 10,774,000 in 1978 (for California the figures are 1,712,000 in 1960 and 915,000 in 1978). Again, the blame is placed on the coyotes even though these ranchers know that it is socioeconomic factors far more potent and deadly than coyotes that are the real cause of this decline.

While this destruction of coyotes is deplorable, it is true that they are adaptable, have expanded their range, and seem in little danger of extinction. It is the destruction of the other species, the "innocent" species, that is especially repugnant. For example, during the years from 1970 to 1977 when 605,000 coyotes were killed, the ADC also killed 36,000 bobcats, 1,450 bears, and 460 mountain lions (and these are the numbers reported by an agency that would just as soon not have it known that they accidentally kill animals other than coyotes). Again, many other deaths were either not located or went unreported. These examples are figures for years after the banning of predacides: figures for earlier years would have been far higher.

This destruction of wildlife was decried for years. The evolving strength of the environmental movement forced the 1972 ban on predacides, though it should be noted that in 1975, the poison sodium cyanide was reregistered for use solely by Federal agents with 26 different restrictions placed on its use. (Sodium cyanide is used in a device called the M-44—the successor of the "getter"—which is a spring-loaded charge of poison with a bait delectable to the coyote placed on the top. The coyote tugs at the bit and gets a burst of poison in the mouth. The bait is also very tempting to foxes.) However, the ban on poisons did not stop the environmentalist thrust. They attacked the use of steel leghold traps as inhumane, and destructive to many species of innocent wildlife. Government agents like to praise the steel leghold trap, saying that innocent animals can be released unharmed. Our personal experience with these traps is that an animal would be fortunate to ever regain use of the paw caught in the trap. Finally, environmentalists attacked the sheep industry itself, which is already receiving substantial public subsidies in the form of wool support and low grazing fees on government lands. They asked why sheep ranchers should receive government assistance for killing the public's wildlife on public lands.

While groups like the Audubon Society, the Sierra Club, and the Defenders of Wildlife attacked on the national level, local groups challenged the predator control programs operating within their own counties. The predator control program is a cooperative program jointly financed by the Federal government and cooperating counties. The agreement to authorize the program is renewable annually, and in our county (Sonoma) it is challenged every year. In these annual debates, the environmentalist arguments range from emotional appeals concerning inhumane methods to the recitation of statistics demonstrating that coyotes cause only minimal damage to the state's sheep population. For example, they quote the government's own statistics showing that coyotes kill only 6% of California's lamb crop each year, and they add that most of these deaths could be avoided by more careful management practices.

One of my friend Jim's favorite stories regarding these hearings involves my Uncle George. George is a sheeprancher; he is invariably a member of the inevitable panel that is formed to examine the county's involvement in the predator control problem. One year he was telling the environmentalist members of the panel about his problems with coyotes, and he invited them to take a drive around his ranch with him so they could see for themselves the magnitude of his problem. After they had completed the tour, George asked them if they had any questions, and one young man turned to him and said, "You should grow soybeans." Every year, our county supervisors renew the contract for predation control with the Federal government.

From experience on our ranch, we have found the arguments presented by both sides to be only partially correct. Our research and observations show the following:

> 1. As soon as coyotes moved into the area, we began suffering heavy losses (15-40% of our lamb crop and 10-15% of our ewes per year.) The statistics cited by the environmentalists are simply that—statistics—and environmentalists more than anyone should know that statistics are often lies. They say nothing concerning actual events in particular times and in particular places. Losing 6% of the California lamb crop may not seem significant (though even that many is quite a few) until one remembers that many of California's sheep are raised in the irrigated pastures of the Central Valley where hardly anything real, including predators, exists. The wilder areas suffer correspondingly heavier losses in order to produce the gross statistic of 6%. Look at the statistics for the wilder states—Nevada loses 29% of its lamb crop to coyotes, Colorado 15%, Utah 12%, and New Mexico 11%.
>
> 2. Marauding coyotes eat very little, if any, of the sheep they kill. Of the dead ewes I have found, the most that has been eaten is a few pounds of the internal organs. Why do they kill them? It is our opinion (and the opinions of other local ranchers and the trapper) that coyotes are much like dogs that get loose around sheep—they just like to chase what runs. This would explain why it is often the healthiest ewe or lamb that is killed, for they are the most likely to run.

It is interesting here to note Barry Lopez's theory regarding predation among wolves. Lopez notes that wolves stare at their prey when they first make contact, and the prey stares back. Following this staring match, a number of options can occur—the wolf leaves, the prey turns around and saunters off, the prey ignores the wolf and returns to its grazing, or the prey runs and the wolf attacks. Becoming prey is a decision by both the wolf and the other animal. Lopez calls this staring encounter "a conversation of death," wherein both animals *choose* how the encounter will end. This "conversation" does not occur with domestic animals, which is why Lopez believes a wolf loose among a flock will destroy twenty or thirty sheep in a killing frenzy. Lopez believes this is because the sheep fail to communicate anything during the "conversation" —resistance, mutual respect, or appropriateness. The wolf initiates a ritual and is met with ignorance. Lopez notes that animals that do not run and do not otherwise display symptoms of sickness or weakness are not usually attacked. I have noticed much the same thing with dogs and sheep. If the sheep do not run and stand their ground with the dog, the dog often does not know what to do and leaves the sheep alone.

Whatever the reasons that coyotes kill, it is our observation that they eat little of their prey, and they never return to a carcass for another meal. There seems little evidence that they kill out of hunger. On their ancestral range (the high plains), the coyote may return to a former kill, may seek whatever food is possible, but not here. Such is always the problem of generalization; it does not explain the reality of specific times and places. Only knowledge and study of a place reveals explanation.

3. The coyotes that live here migrated from the wilder areas to the north and east, where there has been little predator control activity in the past. These are young, healthy coyotes migrating out of areas of overpopulation. They are not refugees.

4. Coyotes are not native to our ranch. Prior to logging, the area supported a climax ecosystem defined by mature redwoods and Douglas fir. It did not support the wildlife nor habitat necessary for the coyote. After consulting the ethnographies of the area and talking with the older ranchers, we concluded that coyotes did not live here until recently.

5. The belief that it is only the old, weak, or injured coyote that kills sheep is wrong. Besides the observation that these are young, healthy animals being killed, this belief presupposes an illogical premise. One can just see a sleek, young coyote sitting on an outcropping overlooking a pasture full of sheep saying to himself, "No, not one of these fat lambs, they are beneath me; I'll go catch something difficult."

After considering these findings, I asked myself again, "Is it worth it to raise sheep here, or should we let the land go wild, or what else can we do with the land? What happens if the coyotes run us and our neighbors out of the sheep business? Looking at what had happened to other ranches of our area, there appeared a number of options.

1. The rancher can switch to running cattle. However, cattle are even less profitable than sheep. They make less efficient use of the grassland and are unable to get into the pockets on the steep hillsides. They cannot overwinter on the grass as sheep can and must be fed expen-

sive hay through the winter. In an area where it takes 1000 acres of land that one owns to raise enough sheep to make the median income, cattle are not profitable. Local ranchers who have switched to cattle do not survive off their ranching income and must conduct logging on their land or work on the outside to supplement their income. In addition, cattle require a lot more labor, better fences, and more capital. Cattle are at least as ecologically offensive as sheep. They trample fragile microhabitats, and in the summer they transform the creeks into feedlot runoff channels.

2. The rancher can go out of business and sell his land. A large neighboring ranch was recently sold to Louisiana-Pacific to be used for timber production. Surely no one can think that any good can come out of more resource land falling into the hands of such multinationals whose motives are not in the best interest of anything—trees, coyotes, or people.

Other local ranches have been sold and divided into smaller parcels—generally 40 acre pieces. These are then sold as recreational or second-home lots, another form of land use that is symptomatic of a disease that benefits no one. While this disease has many forms or manifestations, its cause is our acquiescence to socioecological destruction all around us as long as we can afford our own little escape, our own park, our own unsullied and unused place. Or they become the homesteads for "back to the landers." The latter use has its good and bad points. First, it provides a parcel large enough for people to operate some form of self-sufficient homestead. It provides the land base for a movement that is healthy and produces benefits we are just beginning to see and understand. This movement provides the in-place research station that experiments and works with new forms of architecture and design, new energy sources, sustainable agriculture, the recycling of human and other wastes, watershed rehabilitation, holistic health, and decentralized politics, all intertwined with conscious attempts to improve the self and clarify its correct role within the ecosystem. Yet, too often the land is not used but becomes the private open space for the owner (as is the case with typical residential subdivision). The land is left to go wild. The number of people making even 25% of their income (including growing their own food) off their land is extremely small.

3. Another option is to find some other crop—as the man said, "Grow soybeans." The soils, terrain, and climate of the region severely limit agricultural potential. Recently, a few ranchers have begun planting vineyards, and there is some hope the area will become another quality wine producing area. Another option is reforesting the land to produce commercial timber. However, is there any difference between these practices and raising sheep? Vineyards are sterile systems where little life other than grapes is allowed, and the harvesting of trees is at least as ecologically disruptive as raising sheep. Some might argue this final point, contending that there now exist timber harvesting techniques that are selective and do no great harm to their environs. But even if such methods are practiced in the future and if they do prove unharmful, they should be compared with equally enlightened means of sheep ranching (some of these approaches will be described further on).

4. Finally, there is the explicit goal of many environmentalists and back-to-the-landers to allow the land to go wild, to withdraw it from production. This argument, stated or unstated, pervades much of the debate over coyotes and about land use in general. It is a position I once held, but I have changed my mind. I do not think we can stand up on every occasion and yell save it, preserve it, make it a wilderness area. Though there are definitely times and places where this need be our approach, it cannot be applied to every place, every time.

In the coming decades most arable land will be needed for production, and the prime question becomes not whether to use it or not, but how to use it. In the past we have too simply equated use with exploitation and destruction, and countered with the equally simple solution of "Don't use it."

Whether this need for more productive land comes about or not, it is time we ceased thinking of land as either used (and thus exploited, screwed up, and lost) or preserved (a park). Land can be used to produce *and* to provide open space, wildlife habitat, and recreational opportunity. The English countryside is a proven example. Almost all the land is privately owned and used, yet the public is generally free to use it for hiking or picnicking. When I was there, I often pitched a tent in some field and woke in the morning to the sound of cows grazing outside. The cows did not detract from my "experience," and hopefully my presence did not detract from theirs.

We also need to look at our motives for moving back to the land, for "reinhabiting" it. Reinhabitation does not mean owning a small park while you earn your income and buy your goods outside. To let one's land go wild in this way is to ultimately place a heavier burden on other ecosystems. Other places are to be "used" to produce our goods, and if they are screwed up, well, that's too bad, it's the fault of agribusiness, and we will join the Sierra Club and force politicians to pass laws to prevent them from doing it anymore. Bullshit. Reinhabitation means to inhabit, to use, to use like the native inhabitants once did wisely and reverently, but to use. Reinhabitation should presuppose deriving some sort of land-based livelihood, or at least a more basic economic integration with your place. Otherwise the concept will become another intellectual fad or literary movement. It will have lost touch with the people doing the work. We raise sheep, and try to encourage health and diversity on our land. We still make abstract statements about how other people and corporations should or should not use their land (and such statements will continue to be very necessary), but we also do it ourselves, here, on this place.

Fine, you might say, you guys sound o.k., but its your ranching neighbors we are concerned about. Well, these ranchers are hard-working, knowledgeable, pragmatic, conscientious, and skilled people who, in their own way, care a great deal about their land, their place. They pride themselves on producing quality livestock. They care. There are not many people in our society of whom one could say the same. They also do some dumb and damnable things, but so do we all.

If we are to actually begin applying ecological principles to our lives, then we must recognize that the prime lesson of ecology is that health is roughly equivalent to diversity. To get rid of sheepranchers would be to decrease ecosystem diversity, and by ecosystem I mean the whole system of coyotes, lichen, redwoods, woodpeckers, and people, complete with their potent wishes, dreams, goals, and desires. To deprive this cultural/biological ecosystem of sheepranchers may be as disastrous as losing coyotes.

This is not to defend the destructive practices of these ranchers. But rather than trying to rid ourselves of the ranchers, maybe we should concentrate on these practices and find ways to raise sheep in a healthy fashion. Maybe we should try to talk with these ranchers and show them how they are destroying the land they love, how it is the economic system and not the coyote that is their main worry, and show them actual means of dealing with coyotes (and other land use problems) that allow them to produce sheep while at the same time promoting health on their land.

Controlling the ranchers' abuses could come from on high via government fiat, but another thrust of centralized power is neither necessary nor desirable. Instead we need to reopen channels of dialogue—not simply a dogmatic, theoretical dialogue, but one based on actual experience laced with practical suggestions and language. The need for example is especially strong; we have found it very difficult to get ranchers to listen to just talk, especially talk having to do with the words environment or ecology.

In pursuing such a dialogue, we allowed ourselves to be talked into allowing the government trapper to come onto our land and set some steel leghold traps. We would not have called the trapper in ourselves, as we did not want the traps, and did not want to kill coyotes. However, our neighbor, who is also suffering heavy losses from coyotes, talked with us and convinced us that he felt the traps were absolutely necessary if he was to stay in business. So we told him to bring the trapper out and we would talk it over. We spoke with the trapper and our neighbor for a long time, laying out the reasons we were opposed to predator control (essentially repeating most of the information in this essay), and that we would allow the traps on a provisional basis. If the traps began catching too many innocent animals that could not be released unharmed, we would ask that the traps be removed. Our agreement to the traps was also based on our conclusions about coyotes in this place (e.g., they are not eating their kills, not native to the area, not under severe pressure, etc.). We also spoke with them about alternative, defensive methods of dealing with coyotes that we were trying (detailed below). It is our hope that some of what we said was heard, and that our attempts at dealing with coyotes are observed.

We are not naive enough to believe our dialogue will cause our neighbor or any rancher to change his mind or practices overnight. The change will be slow, and it will be aided by the continued pressure of nasty environmentalists. If nothing else, we leave the dialogue open and offer practical suggestions. It is our belief that the earth (at least as represented in our local ecosystems) is resilient enough to allow the deaths of a few more predators and some more overgrazing in the hope that in the longer run new practices and options can evolve. It leaves the future open to options for a healthy, used place, a place that provides room for ranchers, conservationists, sheep, and coyotes.

One aspect of this future will be new ways of dealing with coyotes. I would like to describe some alternative practices we are trying and considering. Our basic approach is to promote a mosaic pattern of land use on our ranch. This is a pattern whereby use is determined by the landscape. Those areas best suited for raising grass will be used for raising sheep; other areas will be used to grow timber, or left to go wild. These different land uses are interspersed across the ranch giving the appearance of a mosaic of wild and used places. We have withdrawn the entire southern side of our ridge (about 300 acres) from production because it has been heavily logged and seriously overgrazed, has severe erosion problems, and is in need of a long rest. We have reduced the number of sheep run on the ranch by half, and we rotate our pastures to allow one area to lie fallow each year. In the future, we plan to do more cross fencing to withdraw other sensitive areas from grazing, conduct active erosion control methods, and improve our pastures.

As for the coyotes, our tactics have been defensive. First, we tried maintaining a high degree of presence in our pastures (walking about a lot) in the hope that our scent would deter the coyotes. It did not. So we have begun bringing our sheep into the corrals near our home every night. While this is practical on our ranch, it is expensive (we must feed the sheep while in the corrals), time consuming, and the sheep do not make as efficient use of the grass in the farther parts of the pastures. It is not a practical method for ranchers like our neighbor who has sheep spread over thousands of acres of rough terrain.

Other options that we have researched but not yet implemented include:

1. Coyote deterrents. Bells on the neck of the sheep supposedly deter coyotes, but recently our neighbor discovered two dead belled ewes. I think bells might work for a while, then the coyotes learn they are not dangerous and then they become dinner bells. The government is also experimenting with various scents that will deter coyotes, but so far none have proven effective.

2. Coyote-proof fencing. Such fencing may be practical in flat, open areas, but it is not practical in rough terrain where trees fall across the fence and wild pigs make holes through anything that gets in their way. Besides, such fencing is quite expensive.

3. Guard dogs. There are several varieties of dogs (e.g., Hungarian Komondor and the Great Pyrenees) that have been bred for centuries as guard dogs. Traditionally, these dogs lived with the sheep and repelled any predators. They were used in conjunction with a herder so they had constant or frequent human contact. It is questionable if these dogs will stay out with the sheep by themselves for any length of time. Also the dogs work most effectively when the sheep are bunched together where the dog can keep an eye on them, and not spread out over the range as they are here. The dogs are scarce and expensive. Most of them in this country were brought here for show purposes rather than working stock. We are considering these dogs as they could prove effective on our ranch where there are not great numbers of sheep, and where they bunch together at night in a few, predictable spots.

4. Lithium chloride. This is a drug that will make a coyote very sick, but it does not kill. The idea is to lace a carcass or bait with the drug so that the coyotes will eat it, become ill, and become conditioned to stay away from sheep meat (aversive conditioning). The Canadians have had some success with this method, and our government is conducting extensive testing. However, it does not seem it would work here as the coyotes do not return to their kills.

5. Toxic collars. These are plastic collars filled with 1080. When a coyote grabs the throat of the sheep, it gets a mouthful of poison. This is a good idea as it insures you kill only the offending animal. However, it is still experimental as regards the design and the dose of poison. They have found that coyotes learn to avoid those sheep wearing collars and they are too expensive to fit each sheep. Also, one cannot be continually bringing in the sheep to refit small, growing lambs with a bigger collar. Finally, 1080 is not legal and the collars cannot be legally used except for experimental purposes.
   6. Herders. A well-tried method is to provide a permanent herder for the sheep. This is still done in the high plains area where large flocks of sheep are kept together and regularly moved about. However, herders are expensive and practical only in areas where large flocks can be kept bunched and moved. They are not practical in more marginal, rough areas like the Sonoma Coast, where the sheep are spread out over rough terrain.

Being forced to deal with coyotes has taught us a great deal, and will no doubt teach us more. It has particularly made us look again at where we live—its biology, physical components, history, culture, place in the larger biological/sociological system, and people. It has made us grapple with the concepts of use, preservation, health, and need. The coyote, the shape changer, is an apt metaphor for these lessons. In the Indian collections of Coyote tales, sometimes the coyote is portrayed as good, sometimes bad, sometimes intelligent, sometimes dumb—just like the rest of us. And bad and good are relative, changing; the only true fear, true danger, is stasis.

Yet the coyote is more than an interesting or pretty metaphor. She is a concrete animal, too often the easy friend of those with nothing to lose and the declared foe of those who do. She is a very real animal with tangible habits and with whom real people must contend. Decisions regarding the coyote involve a blend of conscious and unconscious reactions to the real and supposed manifestations of an animal who does particular things in specific times and places. These decisions involve our thoughts, feelings, and desires as well as more abstract matters of ecology, economics, politics, and ethics. Our minds and the ecosystem truly are connected. To see the coyote as an actual animal in a place and deal with her so, not as a generalization or symbol, can lead to a greater understanding of our place and our role in it. The coyote/sheep question is a process. It can be viewed as another skirmish in the ecological debate, or as an invitation to participate.

THE RIVER HAD TWO MOUTHS THIS YEAR

Sections from a salmon enhancement Journal *Linn House*

11 September 80

Today we finally have in hand copies of our first-year program design. It's been more work that we had imagined—two months in the making, piling up to more than fifty pages. In it, we have not only shown that king salmon in the Mattole River must be paid special attention if they are not to become extinct, but we have presented an approach to the problem which is, in California, innovative and radical.

Briefly, it goes like this. Eggs must be taken in the Mattole in order to preserve the unique adaptation of the native run. To do this, we'll have to trap fish and hold them until the females are ripe or ready to spill their eggs. Once spawned and fertilized, the eggs will be held in baskets under cold, clean running water until they show an "eye", indicating that they are hardy enough to be moved. At that point, the eggs will be transfered to small, streamside, gravel-filled incubation boxes at various locations in the watershed. The boxes simulate an ideal natural spawning site and provide an egg-to-fry survival rate eight to ten times higher than in nature. The emergent fish will release themselves into feeder creeks that are not currently being used by king salmon. Over the next four years, we will continue to release fish at these sites at the same time repairing whatever it is that made the creeks unavailable in the first place. By the time the adult fish return, we hope to have extended the area available to them for spawning and rearing, thus restoring an increment of natural provision.

Because so little is known of salmon life on the Mattole, we have surveyed the feeder creeks on which we plan to work this year, interviewed long-time residents as to the creeks' historical use by kings, gained a notion of the condition of the habitats, and undertaken rudimentary population surveys. Further, we have provided ourselves with designs for the hardware to be built: trap and weir, a preliminary incubation set-up, the hatch boxes themselves with filters and water supply systems.

The design was in large part the work of Gary Peterson, our biologist/ consultant/friend. Gary did his master's thesis on streamside incubation boxes, working with the California Fisheries Research Unit and Dr. Roger Barnhart. For the last two months he's camped out here near Petrolia, either in his van or in the schoolbus on David Simpson's place. Despairing of our disorganization, ducking our kids' abuse, driven skitterish by poison oak and the problems of steam access, he pushed the design through to completion with a rigorous eye for detail.

17 September 80

The headline on page three in the *Eureka Times-Standard* reads "Board Approves Salmon Project." Since the county has no real authority over its own fish and wildlife concerns, we don't really need the approval of the Board of Supervisors, but we certainly need their support. Armed with a letter of recommendation from the local Grange, we approached the board in chambers which were both plush and sterile, like the inside of a new car. We were received enthusiastically, but the county pleaded poverty and was only able to offer us five hundred dollars worth of feed for whatever fish we decide to rear.

An unanticipated side effect was thirty seconds or so on the local television newscast. Translated through television, our educated guesses on the decline of the salmon runs on the Mattole have acquired a strange authority, and some people listen to us more carefully these days.

21 September 80

John Chambers, Petrolia rancher, invited David Simpson out "to meet a friend from PL." David took me along.

PL is Pacific Lumber, a timber company locally owned until recently. Among loggers (and logging critics, too) PL has a reputation for doing good work, which means taking out the timber but leaving the watershed intact. The friend is Martin Marks, an executive of the company, out for a weekend of hunting.

After looking the proposal over and offering technical suggestions, Marks feels that PL can donate redwood for the incubation boxes and fir for the holding pen. An auspicious Sunday afternoon.

22 September 80

Drove out to Mad River hatchery for a meeting with Ron Ducey, manager of the hatchery, and Dave McLeod, head Department of Fish and Game (DFG) biologist for this area. Ducey is responsible for the artificial incubation of salmon and steelhead eggs throughout the Coast Range. The Redding office of the DFG has assigned him the task of being our advisor and liaison for the department. In California, the DFG is so short-staffed that the recent proliferation of community groups interested in salmon enhancement actually serves to strain its people all the more. Amateurs in the field must after all be trained and/or supervised in the more delicate aspects of fish handling, sorting, and rearing. The art of fertilization, with its close timing, is considered beyond the scope of the self-strained naturalist. While the amateur probably has more potential than the professionals are willing to admit, the responsibility for the well-being of all the salmon and salmon eggs in the Coast Range falls to a handful of right-minded, but overworked, state-employed biologists. Ducey and McLeod could only just spare the two hours they spent with us.

In the past year and a half, David Simpson and I had spent many days in correspondence and meetings with various people in the DFG at the state level, and it had been difficult. The Department had never allowed a local group this much autonomy before, and we had encountered a bureaucratic obstacle course designed to discourage such pursuits. Our tenacity had won us a grudging letter of recommendation from Charles Fullerton, Director of DFG, and this was our first meeting on a local level to work out the point-by-point details. A good meeting today would allow us to begin work this year; a bad one could present a new set of obstacles which could delay us until next year.

Ducey is a large, brusque man whose job at the difficult Mad River hatchery has given him the air of a harrassed social worker whose clients don't want to be helped. McLeod has the dour primness of a good scientist who finds the emotional claims of special interest groups less than adequate. We put our proposal on the table between us, letting it fall from as a great a height as was courteous, and hoped for the best.

The four of us plowed through the proposal and Ducey outlined our relationship to the Department along each step of the way. Hatchbox sites will be inspected by a game warden. The trap will operate only while DFG personnel are present. Sorting of fish (separating males from females and determining when females are ripe) will be done only by Department biologists. Since Sacramento is requiring that our group reimburse DFG people for time spent with our project, it's going to be expensive. We'll have to come up with a salary for something like one person-month or a thousand dollars. We assured them that the money would be there, although we weren't quite sure where it would come from.

Beyond political considerations, they found our proposal rational and we found their detailed advice most helpful. With one snag. Our plan calls for moving the adult fish, once trapped, to an artificial pond where they would be held until they are ripe or ready to spawn. (Gary had gone to no little trouble to make sure the ponds we had in mind were adequate.) The biologists feel that the trauma involved in moving adult fish reduces fertility so much that we would be working at crossed purposes with ourselves. This advice makes it necessary to change our approach: we will beef up the trap so that it can double as a holding pen, leave the fish in the river while they ripen, and take the calculated risk that the fish will be spawned and the trap out of the water before the river rises enough to wipe us out.

26 September 80

Since David was taking his daughter, Gaby, to Davis to begin her first year at the university, he was able to go on to Sacramento for a chat with Assemblyman Bosco. In previous telephone conversations, Bosco had assured us that he could find some state funds for us. By the time David arrived in the capitol, however, Bosco had discovered what we already knew: that state funds for salmon enhancement are meager and overcommitted. The Assemblyman was able to convince the DFG to assist us without reimbursement, and that's a big gain. Mr. Bosco promised to keep our undertaking close to his heart.

29 September 80

Happy meeting this morning with Ed Lewis of the Associated California Loggers (ACL). ACL is a state-wide organization of independent logging contractors. This group has become painfully aware in recent years of the relationship between poor logging practices and the decline of the fisheries, and is anxious to set things right. Lewis has devised a good idea for cooperative efforts between his organization and local salmon enhancement efforts.

Logjams which block salmon migration are a major factor in the decline of the runs. The logs were left in the creeks at a time when their market value wouldn't pay for hauling them out. Twenty and thirty years later, however, the market value has skyrocketed and in some cases a few logs in the jams are still sound. Lewis suggested that we locate jams with roaded access on the creeks we plan to work on. Volunteer ACL loggers might then set up a skyline yarding operation, clear the jam, haul the marketable logs to the mill and turn the proceeds over to us.

It would fall to us to conduct negotiations with the landowners involved, the DFG, the Water Resources Board and so on. There is not enough time to get that done this year, but the idea is a great hole card for coming years.

2 October 80

Two men who trap salmon for DFG came out today to look at the river and to exchange views. Deeply committed to the survival of the native runs, these guys have valuable information to share. They know more about the freshwater behavior of North Coast salmon than anyone we've talked to. Their favored technique for trapping fish is the gillnet, which surprises me. As a fisherman, I'm used to thinking of the gillnet as particularly lethal, seriously damaging the fish it doesn't kill. The trappers have adapted this technique to the capture of live fish by considering the gillnet as an expendable and replaceable tool: as soon as a fish hits the net, it is cut out and transferred to a holding pen. The major advantage of the net is its mobility. You can take it to where the fish are. We were interested and gladly accepted their offer to come out and trap fish for us on their weekends, should our system fail.

31 October 80

Because the river is high and muddy during most of the king salmon run, nobody can be quite sure how long it lasts. Everybody, however, has an opinion. One longtime resident has told us that under the right conditions *all* the salmon can be past in three nights. Another has told us that there are *always* kings in the river right up until Christmas. Biologists shrug their shoulders.

We feel we must be ready for anything, and already it's a week or two past the date of the first possible migration. So David and I have worked for ten days without stopping, building the trap and weir. John, meanwhile, has been putting together a primary incubation trough a few hundred yards upstream. We are lucky to have the help of Jim Groeling, who not only has the best woodworking shop in the valley, but one of the best minds for design and joinery. The shop sits beside the river and the trap is to be installed not a hundred yards away. We are also lucky with the weather. As we have been delayed by one thing and another, so has the rain which will open the mouth of the river held off. But we know we are late, and high tides this week are pounding away at the sand bar which separates the Pacific from the Mattole.

What a flood of details! Should the egg trays sit up off the bottom of the incubation trough? What size and how many bolts will make the trap strong enough to withstand the force of the river but allow for quick dismantling when we want to get it out? What size mesh to hang on the weir to stop the fish coming upstream but not the leaves coming down? What size pipe will allow 15 gallons per minute to flow through the incubation boxes? And where is our next money coming from? My mind is always in two or three places at once and I make mistakes because of it.

But the main structures are almost together and it looks like we may be ready for the salmon when they run.

2 November 80

Nearly an inch of rain yesterday gives today's work party a tangible urgency. The lagoon is swollen and brown. Sea waves wash all the way over the bar. While one crew puts finishing touches on the panels which will make up the trap, another sets up a ten foot ladder in the river and pounds 8-foot steel fence posts into the gravel. The finished panels are wrestled down the bank of the river and bolted together. This action requires workers to be in the river up to their bellies. The only way to keep your clothes dry is to put on waders and take off your shirt. There is no way to stay warm.

The trap, with bolts which have to be fastened underwater, goes together easily: the reward of Jim's careful planning. Next the trap is dismantled and taken out of the river, a rehearsal for the end of the season when the river is sure to be much higher. That too goes easily. But the one-inch mesh on the panels which make up the weir is too small! One hour in the water and they pick up enough leaves and debris to make them into dams. Impossible to clean because the leaves cannot be forced through from the upstream side. The damned thing will have to be redesigned and rebuilt.

7 November 80

Another inch of rain in six hours this morning. David, Bob Bush, and I managed to get the trap into the water yesterday. We spent today pounding more stakes into the river bottom and making numerous trips to the ocean beach 3 miles away to fill sandbags. We have decided to use two-inch chicken-wire for a weir. The sandbags will be used to plug up the holes where the weir doesn't sit properly on the irregular river bottom.

By 3 PM, we were ready to string the wire across the river, but the water had risen a foot and increased tenfold in volume, making it impossible to keep one's footing in the main channel. We retreated to wait for the river to crest and recede, which will probably happen tonight.

8 November 80

Close to midnight last night, there was a 6.5 foot high tide with westerly winds behind it. (The mean high tide is between four and five feet.) At 2:35 AM, we experienced the largest earthquake in recent history, 7.1 on the Richter scale with its epicenter some 40 miles out to sea in the northwest. Our house rocked like a small boat broached to the swell. Sometime before dawn the river bar opened up at its extreme north end, nearly a quarter mile north of where it had opened last year. I don't suppose we'll ever know what influence the earthquake had on the opening. If there aren't salmon in the river now, there will be at the next high tide. I'm feeling small and tired this evening, a minor actor in a large drama.

All day the river remained too high to work in and too muddy to see if there were fish moving.

9 November 80

Despite light showers this morning the river had dropped another six inches. By noon we had the weir strung all the way across the flow and then some, to allow for high water. Two of us spent the afternoon carrying out sandbags, fifty pounds each, one at a time, 60 or 70 of them. Since the fish tend to run at night, we'll take a break and come back at sunset.

10 November 80

Last night made it all worthwhile. With a strong sense of ritual, we went down to the river to greet the fish, one species to another. Around a fire with a willow break between us and the river, we sat and dreamed, several adults and half a dozen pre-teenage kids. It was cold and very still; the only sounds came from the current and the crackling fire. When people spoke, they whispered. The first sign of salmon, a series of sounds like ripe pears being dropped into the river one by one, made everyone catch their breath at once.

All sense of purpose had left me. After all these months of thinking salmon, it was enough to see them. For a moment, I experienced timelessness, human and salmon, co-evolved, strung like a constellation on this North Pacific coast.

Six or eight fish moved slowly and languidly, breaking water but never jumping. They explored the weir without frenzy, moving along the fence methodically searching for a way upstream. The mouth of the trap is situated near the deepest part of the channel, and the trap is placed so that the strongest current is flowing through it. This placement should draw migrating salmon irresistibly into the trap, but in this case, the fish act as if the trap is not there. Each fish edges its way systematically along the weir toward the stronger current and, reaching the mouth of the trap, swims right past it and retreats to hang in the current downstream. Some fish repeat this movement several times.

In order to learn more about the run, and to observe the trap, we have decided to maintain an all night watch, in shifts. Still enthralled, most of us had wandered off by midnight, when John Vargo took over. At about 3 AM, John decided to take a more active approach. Poised on a plank stretched between the trap and the shore, he dip-netted two jacks and transferred them to the trap. (Jacks, or chubs, are two-year-old fertile males.) Toward dawn, Greg Smith came down. Together, Greg holding on to John's belt, they netted a beautiful female which must weigh more than twenty pounds.

This morning, the female still jumps against the lid of the trap from time to time, but for the most part, all three fish are quiet.

11 November 80

River has receded another six inches and the water temperature has dropped from 49 degrees F. to 45.

There are less fish coming up the river tonight and they are still ignoring the mouth of the trap. John managed to net two more jacks.

The mouth of the trap is made of three-quarter inch aluminum conduit, hung vertically on two inch centers. The bars are set up so that they will swing into the trap six inches, but can't swing out. They are unpainted and make an odd metallic clank when they are struck. Maybe they spook the fish. Or maybe, when there are more fish in the river and the spawning urge is more advanced, the salmon will swim into the trap.

12 November 80

The river has turned cold and clean, the opening to the sea is reduced to a trickle. Last night's watch saw no fish, and it is unlikely that any new migrants are coming in from the ocean. We will open up a panel of the weir to allow free passage to whatever fish might be downstream. Decommission the trap, and wait for rain.

19 November 80

We have been suffering drought conditions. On the North Coast, only the Smith and the Klaman rivers show enough water to allow salmon upstream. We finally had an inch of rain last night, which brought the Mattole up a foot. We''ll fish tonight.

20 November 80

A little more rain yesterday afternoon brought the river up a few more inches. At about 11:00 last night, David watched the weir wash out, the steel fenceposts simply laying over in the current. It seems fruitless to rebuild the weir along the same lines at this point, since it didn't withstand this relatively mild storm. And it's too late in the season to try to install something more substantial. It looks like we'll have to turn to the gillnetters.

24 November 80

The rain has stopped, the river has fallen again, and the DFG trappers were tied up over the weekend, unable to come to our aid. It's quite late in the season now, and we're sure that the run will come fast when the water level first allows. We may only have one or two nights to do our work.

This morning Jim Hopelain, the biologist-trapper, drove out with two state-owned gillnets and spent half a day teaching us how to work them. Since the run is over on the Klamath, where he's been working, he'll be able to leave the nets with us for an indeterminate length of time.

The principle of the gillnet is simple. It must be used at night or in murky water so the fish swim into it unaware. The mesh is large enough to allow the fish to get its head and gills through, but not large enough to allow the rest of its body to pass. Trying to back out of the net, its pumping gills prevent escape. Should the fish sense the net before his head is all the way through, its instinct is to turn sharply, catching a mesh with its nose and wrapping itself hopelessly in the net. The net, suspended by corks at the top and weighted with lead at the bottom, must hang loosely in a gentle current. If the net is held taut, it becomes much less effective.

28 November 80

Six sets with the gillnet at the trap site at various times of night and day, all to no avail. The current here is too fast; it is too easy for the salmon to avoid the net. All day today reports kept coming in from Dogleg Hole, two miles downstream, where people are seeing fish. We have spent the day reconsidering whether or not to move live fish. It's hard to see how trucking the fish two miles on a good road could add substantially to their trauma after being gillnetted and handled.

29 November 80

John Lyman had seen fish in Dogleg Hole yesterday too. He lay awake last night trying to figure out how to catch them. One of the more active subsistence fishermen on the lower Mattole, John has spent a good part of this season "giving back to the fish," working with us whenever he had the time.

It was John who figured out how to take the fish in the hole. We would set one net at the downstream end and walk the second net toward it, starting a hundred yards upstream. It took us an hour to get up a crew of six and to put an 1100 gallon plastic tank half-filled with Mill Creek water on the back of Danny Rathbun's flatbed truck. It's become obvious that we're not going to get our eggs this year unless we move fish. It seems worth the risk.

When we arrived at the hole, we could see the bright shapes of large fish in the murky water under the willows. Considering the thrill factor, we got out nets in the water with surprising grace. The crowding net hadn't been moved downstream 30 yards before the first fish hit the set net. John and I were in a boat between the two nets and as I pulled the fish into my lap to cut it out of the net, I was peripherally aware of more fish hitting both nets. Mind stopped working and adrenalin took over. Mind stopped working to the degree that I couldn't tell what kind of fish I had in my lap. It was a steelhead, and steelhead weren't supposed to be running for several weeks yet.

"John, what kind of a fish is this?" I yelled.

"Why, it's a salmon, Linn."

"No, it's a goddamned steelhead, John. Look at that white mouth." With that, I threw back into the river a fat and vigorous steelhead which surely weighed fifteen pounds, a fish which to my palate is the best eating of all the salmonids. What painful altruism. John turned gray. The fish released swam right back into the net, and so it went for half an hour, all of us working feverishly with much shouting and splashing, cutting fish out of nets, sorting them as best we could, and hauling the obvious salmon in dipnets up the bank to the tank truck. After handling fifteen or twenty fish, some of them several times, we took back to the trap three king salmon, two of them female, a pair of silver salmon, and one steelhead, which we released. The fish took the trip well. Settling into the tank quietly, they showed as much vigor when we transferred them to the trap as they had in the river.

David, with his usual abandon, had jumped into the river fully clothed to cut fish out of the nets. Once the fish were safe, he got into a car soaking wet for a two hour ride to Garberville, where he's scheduled to make a speech at a fundraising event. I hope he doesn't catch pneumonia.

2 December 80

We have spent the last three days searching all the holes in the lower five miles of the river, looking to repeat our success on Saturday. When we could get a crew and the tank truck together, we've fished the pools. Otherwise, we've had the nets in at the trap site. We've seen no more fish.

Last night in a heavy cold drizzle, David and I made a set at the trap site. Discouraged and uncomfortable, we pulled the net about an hour after dark. Some sort of prescience caused us to move an outboard motor and other valuables up off the river bank. A short time later, it really began to come down. I woke up a dozen times during the night, listening to the downpour that went on and on, and at one point well on toward dawn, I remember thinking, "That's it, we're done."

By 6 AM, it had rained five inches. The trap, with our precious salmon in it, was two and a half feet under water and twenty feet from the new shore. A rapids had formed over the rooted snag which holds the trap in place. Oddly enough, our first concern was not for the time and materials we had poured into the trap, but with the safety of the ten fish within it. It was too painful to think that we might end up killing the fish we were trying to save. Ten years ago, I might have risked going out there to break up the trap. This morning, I was agonizingly aware that to do so would be pressing my luck a little more than I was willing. By 9 AM, we'd had another inch of rain and the trap was gone. John Vargo, who doesn't practice loose talk, reported seeing pieces of the trap floating away from the site. So we can assume the fish are free in the river.

There is a massive new opening in the bar at the mouth of the river, straight out to sea from the main force of the flow.

22 January 80

It was a good year to take our first licks at the complexities of salmon restoration. between drought, flood and earthquake we've had a good initiation into the extremities this watershed has to offer. We know what to build for next year.

In the weeks following the unseasonal flood of 2 December, we made several attempts at reconstructing the year's effort. A new holding pen was installed fifteen miles upstream and we readied ourselves to follow the salmon up. The trappers came out to get their nets back and revealed that DFG people in Redding were deeply concerned that we had possession of the nets at all. A question of equity. There followed a series of lengthy and sometimes emotional telephone conversations between our people and various Department officials. The Department took the position that we weren't to pursue the fish onto the spawning grounds, that anything we did upstream had too much chance of interferring with natural processes. Since sport fishermen were allowed to fish as far upstream as Honeydew, there was room for contention, but overall, it was a difficult argument to refute and the decision stood. Coincidentally, our funds were depleted and some of our workers were uncomfortably in debt.

While our efforts downstream were concentrated on capturing fish, other groups upstream had been building and installing incubation boxes to receive the eggs they hoped would be forthcoming. DFG was alert to our need to test the boxes this year and would have provided us with silver salmon eggs. But all the hatcheries in the Coast Range had suffered the same difficulties that we experienced this season, and there simply were not enough salmon eggs to go around. At this point it is very likely that we will receive 30,000 steelhead eggs in March, so that the hatchboxes will be tested in time for the next season's undertaking.

Any hatchery operator can tell you that capturing live salmon is difficult. The degraded watersheds in the Coast Range, with their ever more rapidly fluctuating cycles of flood and drought, increase the difficulty. But as long as there are native strains running in the rivers of the Coast Range, it remains an intelligent undertaking to do everything in one's power to preserve this miracle of adaptation. Ten or twenty years of work seems a light enough price to pay for the continuity of 30,000 years of genetic diversity and natural provision.

There are a couple of things we'll do differently this year. We'll need to construct sturdy, semi-permanent stanchions in the river on which to hang a weir. We need to reconsider the mouth of the trap. We should go back to our original plan of a secondary holding pen up off the floodplain.

Local support for the program has grown this year, and we have every reason to believe that with support from responsible and thoughtful people elsewhere it will continue to grow. And for as long as we believe that we **can** make the repairs necessary to the places in which we live, there will be, for some of us, no work more compelling and reasonable.

A NOTE ON FUNDING

When we first began pursuing this effort a couple of years ago, it seemed that once we were able to convince the state to let us do the work, we would surely find state funds to support it. We not only didn't find those funds, but discovered in the process that the state's appaling inability to do anything about the conditions we described made the kind of citizen effort we envisioned more of a grinding necessity than an interesting alternative. Later, we found that the lack of state support had its benefits. Most people, both as volunteers and as financial contributors, enjoyed the sense of direct action and mutual endeavor that our project gave them. This is not to say that we would categorically refuse state aid in the future, but that we have discovered a considerable set of values outside it.

With the exception of $1250 from private foundations, all of our working monies came from individual donations. To our surprise and gratification, much of that support came from other watersheds and distant cities. We are glad to assume that these contributions mean that our work might have some universal sensibility and application.

Last year's work on the Mattole amounted to a full time job for four months for two people and probably that many hours again in the combined efforts of other people in the valley. Those kind of hours cannot be sustained on a volunteer basis. There aren't enough hours in the year for the individuals involved to subsidize a third of their time in the marginal economies which barely support us all.

In order to sustain, improve, and enlarge on the work begun last year we will need to raise considerably more than the four thousand dollars which has taken us this far. Over the years, it is not immoderate to project an annual budget of twenty to forth thousand dollars. Responsible logging and commercial fishing organizations are beginning to come around, but there can be no doubt that the immediate future of this effort remains in the hands of concerned individuals. Our address is Mattole Watershed Salmon Support Group, Box 189, Petrolia, CA 95558.

Joel Beck

COMICS   Mort McDonald

INTERVIEW WITH A GENTLEMAN FARMER

Bruce Boston

Mr. Dixon's farm occupies 130 acres of Northern California countryside, where in springtime streams abound and hillsides are lush. He first took up his profession nearly thirty years ago, after returning from World War II where he served in the Pacific under MacArthur and was thrice-decorated. Today, many of his competitors consider him one of the finest breeders in the business.

I arrived at the Dixon farmhouse, a salmon pink ranchstyle, early in the afternoon. Mrs. Dixon, a plumpish merry-faced woman, met me at the door, wiping soapy hands upon her apron. She informed me that her husband was "out back a'workin." As she led me through the livingroom and kitchen, which were modestly, even severely, furnished, I asked her a number of questions without success.

"I really can't tell you a thing," she pleaded, "I leave all the gentlemen to my husband."

To the rear of the house there were several other buildings, including an immense barn, bright red and crisscrossed by strips of white planking. Beyond the gently rolling fields, green hills rose up on every side. In the midst of this picture postcard setting, Mr. Dixon emerged from an open doorway, wrestling with a large bale of chicken wire. As we approached he took off his hat, a simple blue baseball cap, and with one forearm wiped the sweat from his brow. He was a tall, large-boned man with narrow eyes. His salt and pepper hair, barely receding, was crewed closely to the skull.

We exchanged greetings. Since he had "a good deal a'work to do," he suggested we get right down to the interview. "Might just as well stretch our legs a bit while we're at it," he added. Screwing his cap back on he took off at a brisk pace. I fumbled pad and pencil from my coat pocket and went after him. Mrs. Dixon, seemingly unnoticed by her husband, turned back to the house.

"What impact have recent economic trends had on your work?" I began.

"They're no skin off my nose," Mr. Dixon informed me. "If you're a gentleman farmer you never have to worry." He nodded curtly. "There'll always be a demand for good gentlemen."

I saw a small herd of gentlemen, perhaps twenty, grouped at the far end of the field we were passing. I noticed that they were all naked.

"A true gentleman is always a gentleman whether he has any clothes on or not," Mr. Dixon explained. "I remember a few years back, one of them larger concerns . . . the combines, we call them . . . brought in some Australian stock. Tried to flood the market, drive the prices down so they could force fellas like me out of business." Mr. Dixon stopped abruptly and turned to face me. His eyes narrowed still further, glinting silvery in the afternoon sun; then he broke into rickety laughter and slapped his knee. "They dressed 'em up in silk suits, cravats, fancy boots, the whole shebang! Then they had a show over in Chico of all places. Made complete fools of theirselves, that's what they did. Some gentlemen! Why those damn Aussie's didn't even know which fork to use on their salads." He laughed again, dabbing tears from his eyes with a checkered bandana.

I asked if I might take a closer look at some of the herd, and Mr. Dixon proceeded to instruct me in the proper method of summoning gentlemen. Taking out his wallet he removed several bills and began rubbing them between thumb and forefinger. "Money," he called, his voice rising yodel-like on the second syllable, *"Mon-ey! Mon-ey! Money!"*

The gentlemen came bounding across the field in groups of twos or threes to gather on the opposite side of the fence. Mr. Dixon dug into his pocket and came up with a handful of coins which he tossed into their midst. As the gentlemen scrambled for them, he pointed out his fields to me. "Pure dichondera," he said, "I raise all of my gentlemen on beefsteak and dichondera. When people complain to me about the price of gentlemen, I say, 'Look at the price of beefsteak. Look at the price of dichondera. One thing you can be sure of. When you deals with me, you gets what you pay for.'"

By this time the gentlemen had settled down and were eyeing us expectantly. They were all young bucks and looked like a healthy lot, but I also noticed that they were quite pale. I asked Mr. Dixon about this.

"Pure Anglo stock," he beamed. "No wops or spics for me. I can only afford to let them out for a couple of hours each day. A true gentleman will always burn before he tans."

Mr. Dixon then proceeded to engage his gentlemen in conversation. In the space of a few minutes they covered a wide variety of subjects. Not only salad forks, but entering and exiting elevators, the fine points of hat tipping, tracking lions at the Bronx Zoo, and tea bags versus moustache cups as methods of imitating the Queen. Mr. Dixon was right about one thing. The carriage, manner and enunciation of his stock were so uniformly flawless that I soon forgot they were without clothes.

Mr. Dixon glanced at his watch and quickly called the conversation to a halt. He took me by the arm and began guiding me back in the direction we had come. "I imagine you'll want a look at one of the breeding pens before you go," he said. I had hoped for a much longer interview, but I could see that I was confronting a man truly committed to his work—in his own way, perhaps something of an artist—and I felt thankful for the time which I had been allotted. Mr. Dixon's gentlemen followed us along the fence line until they reached the perimeter of their enclosure.

"Good day, sirs," they called out in a ragged, yet hardy and melodious chorus.

With amazing agility for a man of his age, Mr. Dixon clambered up the side of a concrete bunker, painted the same bright cherry red as the barn. He reached down a hand and hoisted me up next to him. We walked across the roof to a wooden trapdoor. First sliding a bolt, Mr. Dixon lifted the door by means of an iron ring embedded in its surface. It fell back heavily against the roof.

A shaft of sunlight shot into the depths of the pit below us. I glimpsed a confused tangle of white limbs, shoulders, patches of hair, before they shifted quickly back away from the light into darkness, leaving only a small rectangle of straw-strewn dirt. The sounds of grunting, gasping and cooing issued forth from within. And other sounds, which I could only liken to the rattle of chains and the quick slap of leather upon flesh. Yet most striking of all was the odor which assailed my nostrils: musky, pungent, brothel-like in the extreme. I leaned back away from the opening and gulped fresh air. Mr. Dixon shook his head.

"Although a gentleman is the cleanest of all God's creatures," he proclaimed, "he likes his sex dirty. Lots of gentleman farmers don't seem to understand that, and that's why they fail. You have to have a stomach for work like mine." He crinkled his nose as he leaned forward and let the door fall shut. "But even I can't take that stink for very long."

Mr. Dixon accompanied me to my car. As I thanked him for the interview, he placed a hand upon my shoulder.

"My pleasure, young man, my pleasure." His hand slipped lower, kneading the bicep. "You're a sturdy fellow, aren't you?" He eyed me speculatively. "You know, it's never too late to think about becoming a gentleman. What did you say your ancestry was?"

I assured him that my mother was a full-blooded Magyar Jewess, slammed the car door on his fingers, and as he hopped off across the fields yelling, hastily made good my departure.

PRIVATE BRANCH EXCHANGE

Lucia Berlin

It was dark. Phyllis, Operator Five, waited for the 6:32 AM bus. She had done this for 23 years now and it had not occurred to her that it was just as dangerously dark in the morning as it was at night. She never went out after sundown unless in a taxi . . . and then she took a can of hairspray in case of attack. Attack by whom? Coloreds. Bloods, they called themselves, on the bus. Blacks. Her throat closed on the word black, although she knew that was what you were supposed to call them now. Even the word was too dark.

Jesus Mary Joseph. She said it out loud. She still couldn't believe it . . . a colored supervisor at the switchboard. At Hamilton Memorial Hospital, in Oakland. The operators had all been shocked by the decision. They had known Phoebe was retiring, after 40 years, but had taken it for granted that one of them would be chosen to replace her. Not Mae, Operator two, she only liked working graveyard, but all the other AM operators had applied. Most of them had been there for over twenty years.

Thelma (One) had just laughed when they heard. "Well, girls, we're all much too old!" but the others had scribbled notes to each other like "Writing on the wall!" and "Phobe's Revenge!"

That made sense to Five. After 40 years at a thankless job Phobe would go and hire some sassy young colored thing, just so everything would fall apart after she was gone.

At shift change Laura, from Admissions, had come down with roses and champagne for Elaine, the new supervisor. The PM operators had come in early, with roses too, and were chatting and happy. Well, they would be. They were all talking gaily as the AM operators walked stiff-legged down the hall to the time-clock. "Right on!" Elaine was saying, and laughing her throaty, colored laugh. She didn't talk colored though. Like Four said, she sounded just like Jackie Onassis.

Five actually thought of black people as green. Ever since Medi-cal began the operators had to pencil green all the Medi-Cal patients in their rands. Medi-Cal patients couldn't make toll calls, only Collect or Bill-To. Daily, Five made her own private tally of how many greens there were in the OB ward and not delivered. "You just know they're all abortions." They got phone calls all day and most of them went by several names. "Lemme talk to Marie Loutre, or maybe she's under Marie White."

"The greens aren't automatically black, Five," Three tried to tell her, "They're just poor."

"Just trying to survive." Thelma said. Thelma was Operator One, but they called her by her name, even though on the phone she said "This is your Operator One."

Three had made fun of Five when it began to seem like all of the PM operators being hired were black. "You should be happy . . . they're trying to get off welfare and save your taxpaying money!"

"Now, girls," Thelma said. "They're sweet, peppy young things. You were new at the board once too." Thelma was always like that, always a good word. A pushover, really. She did get mad at Dr. Strand. Who didn't? He was so mean, said things like "What do they do at that switchboard, hire the handicapped?" "Thank you, Sir," Thelma said, and disconnected him on a London call.

● ● ● ●

Thelma was the only one plugged in when Five got to the basement. The others were all standing around the time clock, looking like there had been an accident. Five knew they were talking about Elaine or the PMs. Mae, (Two) the graveyard operator, was still there. She was enormously fat, distorted since she had had both breasts removed. Not cancer, just because they were so heavy, 15 pounds each. Three and Four were almost as fat and, like her, had never married. The Mob, Five called them to herself, for Morbidly Obese Biddies. She and Thelma had discussed why there were so many very fat switchboard operators. Probably because no one would hire them if people could see them. Thelma said that they all had perfectly lovely voices. And pretty hands. Most stout people have very pretty hands. Stout? Jesus Mary Joseph.

They took up their positions at the board, hooking on their earplugs and mouthpieces, plugging in. Thelma and Three sat at the page positions. The tiny cubicle was already hot, smelled of perspired polyester, deodorant, Four's cloying Jean Naté. There was barely room for the switchboards and Elaine's desk. The vent was open now but since the air came from the parking lot they would have to close it soon when the fumes got bad.

The early morning hours were pleasant, the board relatively quiet. The women filed cards in the rands above them, admitting and discharging, changing rooms, moving patients from ICU to the floors, green-penciling. When patients died they circled their names in red, paper clipped them to the expiration list at each position. They didn't tell callers the patient had expired but said "One Moment, Please." and connected the caller to the nursing supervisor. Thelma liked to say that her own dying words would be "One Moment, please!" Five said hers would be "Please hold, God."

The AM operators had known each other for many years but never saw each other outside of the little room. They took their breaks separately and parted at the time clock. Except for Thelma, who told everything, they shared little of their personal lives. Their conversation revolved around the switchboard, TV, food and the Doctors. They followed the Doctor's lives in the society pages and the Hamilton Hospital Newsletter. They spoke about them as if they knew them, had "known" them for many years, back when many of them had been residents and interns.

"Dr. Hanson and Dr. Angeli flew with their wives to Miami this morning, then they'll be cruising in the Carribean."

"Carríbean." Five muttered.

Three sighed . . . "That's what I want to do . . . go on a Love Boat!"

Four nudged Five ". . . They'd have to give her a whole life raft instead of a life preserver . . ."

"Look who's talking!" Five wrote to Thelma.

They talked all day long, even though they were very busy. In between calls they talked to themselves, murmuring like Thelma's "I've heard everything now" Three's "Nit-Wit!" and Five's "Jesus Mary Joseph!" Each had her own tune she sang as she filed. Three's was "Old Buttermilk Sky" and Four drove them crazy with "My Foolish Heart" all day. The little room hummed with the different rhythms as, like milkmaids each leaned into her own board, crooning as if to her favorite cow.

If you stood behind them as they worked . . . unless you were as sharp-eared as Elaine . . . it was hard to tell when they were on calls, or who was gossiping to whom.

OPERATOR ONE

Good Morning, Hamilton Hospital. Operator One, May I help you? Dr. SCHMITZ PLEASE, DR. SCHMITZ. RESPIRATORY THERAPY 202. RESPIRATORY THERAPY 202. Room 3201? I'll connect you. Too hot for this girdle. I'm fine, Dr. Miller, thank

OPERATOR FIVE

I'll bet she won't be in before nine. One moment please. We have no room 100. The LAST name please. Ignorant fools. Good Morning, Hamilton Hospital. Thank You. Sorry the babies are in the rooms.

you. Do you want your service, sir? He's so sweet. Wouldn't think it to look at him. The busiest and best surgeons are the most patient on the telephone. Good Morning. Hamilton Hospital. DR. ANDERSON. DR. ANDERSON PLEASE. DR. Wilson? Emergency wants you, sir. MRS. SCOTT PLEASE MRS. SCOTT. Tell me to page, don't just write it down . . .

OPERATOR THREE

Good Morning, Hamilton Hospital. I know what S and M means, but what is B and D? Operator Three. Yes you may, dear. MISS ALBRIGHT PLEASE MISS ALBRIGHT. Is Albright in yet? Sorry she's not answering page. Hamilton Hospital. And she'll be out of here by 3, you just watch. Operator Four, what number please? I've never seen him but Two has and he's white. That line is busy. Call again, please. DR. MILLER SURGERY DR. MILLER. Not light. White. Mrs. Scott? Yes, an expiration call. Dr. Miller? Surgery wants you. Hamilton Hospital, may I help you?

Jesus Mary Joseph. How did they make the babies in the first place . . . they talk the livelong day. Have you seen her husband? One moment sir. Who wants Dr. Winthrop? He's light, you say? . . . Ruben, Ruben, I've been thinking, what a grand world this would be, if the . . . Hamilton Hospital. White? You're kidding. Did you page Mrs. Scott? Operator Five, may I help you?

OPERATOR FOUR

Lord knows. I can't remember what PBX means either. . . . the night . . . is like a lovely tune . . . beware, my foolish heart . . . No expirations today, funny how they come in threes. Good Morning, Hamilton Hospital. Sorry, that party can't receive calls. Operator Four, thank you. Dr. Miller Stat for surgery . . . There's a line between love and fascination that's hard to see on an evening such as this, and they both have the very same sensation when you're locked in the magic of a kiss. His lips are much too close to mine. Beware, my foolish heart!

"Good Morning."

It was the new girl, Operator Nine. (Nine was Mickey's number; she died last year.) Because of the shock of Elaine's promotion the operator's hadn't had a chance to react to Nine yet. Not a girl, really, in her thirties, white at least, with four children.

"You shouldn't clock in so early. Never more than five minutes."

"I wasn't trying for overtime, just didn't feel like hanging around."

"They don't like you to punch in early."

"OK."

"Never say OK on calls," Five said.

"I'll train her, honey," Thelma smiled. "Besides she's real polite on calls. How are you today, dear? Tired?"

"I was! I went to bed at 8 o'clock. Shall I listen to you or take calls, Thelma?"

"It's time to start breaks. Four can take the page. You sit at her board till I get back. Four you watch out for Nine now." Thelma unplugged, straightened her red wig and walked slowly down the hall toward the elevator.

"She's not as strong as she used to be."

"Well, she couldn't be faster at the board."
"She gets awful tired around one o'clock."
"So do I!"
"It's just right with five of us. Look, all the boards are full. It's too hard to take breaks with just four." And as they talked their fingers were flying to answer the lights, pull down the disconnects."

"Test your calls before you pull them down." Five hissed at Nine.

"Aren't those disconnects?" Nine knew the others just pulled them down but said ok.

"You're doing just fine," Four told her. Nine was learning fast, was courteous on calls, not flip or sultry like the PMs. But she'd probably get that way when she started working with them . . . joking with callers, making personal calls. Five hated Nine already. Four knew why. Five had asked Nine why she had chosen to work as an operator. Nine had said "It's the only job I could get." Elaine had laughed and said "I hear *that*."

Neither of them realized how that made the AM operators feel. Like fools, that's how. They had spent their whole lives being good operators.

Thelma talked to Nine about what it was like, when she was a girl, how PBX was one of the few nice professions. "I was proud as punch to be a Hello Girl. Some folks say it's a thankless job, especially in a hospital. Just remember you are often the first contact people have with the hospital. That's an important role. Also remember that people are under awful stress . . . sometimes life and death! When they get mad it's not personal. Somedays it seems like everybody's mad at you. But they thank you all day long, too, for helping on calls, or finding somebody or just sounding friendly. Just remember, it's good clean work, honey," Thelma patted her arm, "You'll make a fine operator, I can just tell."

• • • •

On her first day, when Elaine told her it was time for lunch, Nine had reached up and started pulling down all her lines. The other operators had yelped and screamed.

"You're disconnecting all the calls!"

"You fool! Jesus Mary Joseph!"

Nine was embarassed. "I'm sorry. I don't know what came over me . . . guess I was just tidying up my board before I went to lunch. How stupid."

Elaine and Thelma were the only ones who thought it was funny. "That's the spirit, Nine . . . Time for lunch, folks!" and Thelma said "They all talk too much anyhow. It was a natural mistake, dear." Five had scrawled "Naturally *stupid*." and Four had written "Not as smart as she thinks she is."

Elaine followed Nine to the elevator, still chuckling.

"Nine, tell me one thing . . . "

"Did I lie about PBX experience? Yes, I never saw a switchboard before. Please let me stay. I'll learn it."

"You're doing great. Don't worry about it. This is *my* first day too, remember. And they hate me."

They stood in line for sandwiches and coffee, sat by a window table.

"You'll like it," Elaine said, "It's fun and it's a good skill to know. You couldn't have a better teacher than Thelma. Just don't tell anybody anything and don't let the others get to you . . . petty bitchy bigots. You'd better get back. Don't tell them you were with me. And never get back a minute late, or a minute early."

• • • •

"Elaine, I know I took my break already but I need to go to the restroom."
"Go ahead, Thelma."
When she got back Elaine stood over her board.
"How long have you been working here?"
"38 years."
"You don't need my permission to leave the board. You're number one, in every sense. You can take as long as you like at lunch and can make personal calls. Whatever you say at this board goes. Does everyone understand?"
Thelma had tears in her eyes. She nodded, blushing.
"Now that was a sweet thing to do" Four said.
Five fumed. Sneaky, that's what Elaine was. She knew Thelma would never take extra breaks or long lunches. As for calls . . . Bud called her every day at 2 and all she ever said was "I love you, too" and got off the call. But she'd never side against Elaine now.

• • • •

"Do you all see this button? Just to the right of your split key?"
"I've always wondered what that was . . . "
"It's called a peg button. I want you to hit it every time you answer a call. Inside or outside call. It will be a nuisance at first but after awhile you'll do it automatically. Nine, you're lucky, you can get in the habit from the start."
Elaine stood behind them for several days, reminding them to hit the key. Three and Four were angry. "Is she trying to check up on us? Doesn't think we answer enough calls?" and whenever Elaine left the room the operators hit the peg key dozens of times. Elaine caught Five doing this.
"Listen, ladies . . . I need an accurate count, so I can make an accurate report. You defeat yourselves, not me, if you peg too little or too much. Too few calls will mean we have too many operators. Too many will prove that it's time for Centrex."
"Centrex!"
"Centrex!"
They were silent. You could hear calls beeping from their headsets. Elaine was standing above them, so none of them wrote notes.
At lunch Nine asked Elaine "What's Centrex . . . hoof and mouth disease?"
"Computers. It would get rid of most of them."
"Will it happen?"
"Not for years. Just keeping them on their toes."

• • • •

Friday. Six and Seven would be there until Monday. Three's days off were Thursdays and Fridays, Four's were Fridays and Saturdays.
Six had had a face lift and lived in Walnut Creek. She had left the switchboard years before to marry a wealthy man, had had to return when he suffered a stroke. She never let them forget that she had seen better circumstances.
Three liked to say things like "You know that Miz Suburbia doesn't get paid to work here . . . she's a volunteer, like a pink lady, just comes down here to bring cheer and goodwill to us underpriviledged."

Six drank. They could smell it over the Clorets and many days she was bleary-eyed and shakey. One morning Elaine gave her a Valium, told her to go lie down for awhile. When Six left, Elaine stood behind them so no one would say anything. Six looked better when she got back, smelling of soap and Listerine.

"Thanks, Elaine. I was up all night with my poor husband."

" . . . with poor old Grandad, more likely," Five whispered. Seven was very young and new to the city. How she got hired at all was a mystery to the AMs. Six said it was because the hospital was in trouble for not hiring enough blacks. Seven said things like "Dr. Miller be's in surgery" and "I'll axe my supervisor." (Six said 'I'd like to axe the supervisor' and the girls giggled for hours.) But Seven worked harder than all of them, even Thelma, and was never late or absent. "Seven you are a perfect little thing," Thelma said.

But still . . . to get promoted to AMs. . . . She told them she really appreciated it, that Elaine had helped her out because she had three little kids and none to help her out. Well, Nine and Twelve had little kids too.

Elaine heard that. "And they have friends and family to count on. Are you questioning my decision? That is insubordination."

• • • •

Five liked Fridays because she got to be second page. It made the day go much faster and gave her a sense of importance, knowing that her voice echoed all over the hospital, even in the rest-rooms. Phoebe had told her that she sounded like Katharine Hepburn. Five thought she spoke clearly and crisply when in fact her paging was an embarrassment to the other operators, shrill and jarred together, like a record at the wrong speed. No one had ever said anything though, to her face, and she was third in seniority. (The Original 4's and 5's had died, long ago.)

"I'm sorry, sir. I didn't mean to disconnect you. May I try again?"

"Never cop to anything, Nine," Elaine said. "And never, ever, say you are sorry. It makes them madder. If you have to say anything say "Trouble on the line. Sir.'"

"That's true. Whatever you do, though, don't sass them back. Just disconnect them and count to ten."

• • • •

"Did you read Robin Orr's column last night?"

"About Dr. Steinberg's party? Wasn't that a scream? All the heart surgeons were there. Lot of pull and him so young. No wife, at least they didn't mention one."

"No, just his wolfhounds!"

"Can you imagine giving a birthday party for your dog?"

"And the guests brought their pets."

"Where does he live?"

"Didn't say."

"Yes it did. It said ' . . . a cool Moraga evening. . . .'"

"Dermatologist? I still can't place him."

"We never get any calls for him, but he calls out all the time. From the nurse's station at 2 east."

"He's real tall. Doesn't look Jewish at all."

"He talks real Jewish."

"That's New York, not necessarily Jewish."

"I still can't place him. Does he eat in the cafeteria?"

"Yes, but with the nurses, not the doctors."

"You know, the velour one with chains. Wears Blue Jeans."

"That one! Looks fruity to me . . ."

"It's pretty weird, a party for your dog."

"Well, he can do it because he's rich. If you're poor you're crazy, if you're rich you're witty."

"Eccentric."

"He's eccentric all right," Elaine said from her desk. "His bedroom has mirrors on the whole ceiling and a round bed covered in mink. All over the walls are huge photos of animals, uh, mating. Elephants, pigs, hippos, giraffes . . ."

"Giraffes?" Nine was fascinated.

"Well, maybe no giraffes, but everything else."

"You were there, Elaine?"

"You went to the party?" The operators were stunned. Three and Four scribbled notes. Elaine was on the phone, put her hand over the mouthpiece, "It's the 70's, ladies. We eat at lunchcounters too."

Nine could tell that it wasn't Elaine's color that shocked the operators, but a deeper PBX taboo. A real witness took the pleasure out of their speculations. Even though they knew each doctor's quirks and schedules and as much of his private life as they could piece together they never spoke to any of the doctors "in real life" as Four called it. They didn't ask Elaine who was there, or what people wore. In fact, they said no more about the party.

• • • •

Nine worked now from one to nine at night. She liked the shift. Her mornings were free and her sons still up when she got home. It took care of the evening hours, long now that Terry was dead. He had always hated evenings and Sundays. Nine didn't talk about him, not even with Elaine or Twelve.

Operator Twelve came on at two, shimmering into the afternoon stupor with her easy Italian laugh, 51 bus stories, salami and jokes from her father's Alameda bar. She plopped down at the last position but Elaine told her to take over Five's page.

"I'll page until shift change," Five said curtly.

"You'll file dismissals. You're tired."

Five moved with much slamming and slapping of her things. Twelve plugged in, grinned down the board at Thelma . . . "Hey, One, what you get when you cross a computer with a whore?" Thelma shook her head. "A fuckin' know-it-all."

"Honey, you just missed this man I kept getting on an intercept. He said "Sorry, operator, I keep putting my finger in the wrong hole. Story of my life!"

Elaine laughed from her desk. "Thelma, I can always tell when it's close to three. Your wig gets crooked and you start talking dirty."

"You can't imagine where my wig was when I woke up this morning! I says to Bud . . . honey, you used to make my hair curl . . . now you make me blow my top!"

"Wish I had me a husband like your Bud," Twelve said.

"Me too," Nine sighed. Seven began to warble Four's Beware, My Foolish Heart and all of the operators laughed.

Five could hear Twelve talking, but not Nine's answers, as Elaine, Thelma and Seven were babbling away. Six kept to herself, as always. Too good for us all, the old sot.

Twelve's rich warm voice reverberated from the speakers:

> DR. FIRESTONE. DR. FIRESTONE, PLEASE. RESPIRATORY THERAPY, TWO EAST. RESPIRATORY THERAPY, TWO EAST. Operator Twelve, may I help you? Yes, Dr. Strand, there are several messages. ICU, CCU and Recovery. Also outside calls from Linda, Becky and Samantha. Three from Samantha. . . Sure enough, the sucker asks for an outside line. First things first, right? Good afternoon, Hamilton Hospital. Thank you. DR. ATHERTON PLEASE. DR. ATHERTON. Nine, I told you about Mario hitting on me at the party, but I thought he was just fooling around, for old times sake, and his old lady there and all. MRS. CILANTRO. MRS. CILANTRO. But last night there he was at my front door. Whoa! I almost died. MRS. CILANTRO. MRS. CILANTRO, *PLEASE*. Good afternoon, Hamilton Hospital. One moment. Yes, Mrs. Cilantro, Admissions is looking for you. Operator Twelve . . . sorry Dr. Strand is on an outside call. ... And he kissed me, slow, real real soft but he was holding me hard, you know? My head in both hands. Christ. MRS. CILANTRO, MRS. CILANTRO. DR. WILHELM, PLEASE. DR. WILHELM. I *Know* you got that page, Cilantro . . . Three East wants you now. You're welcome. . . . Bitch. DR STEINBERG, DR. STEINBERG. Did you read about his dog party? We just sort of melted down onto the couch . . . but Luke was asleep on it. Room 5812? Thank you. DR. OVERTON. DR. OVERTON. Dr. Strand? ICU *really* wants you, sir. Yes, sir... He says he's well aware of that fact and give him another line. Well, la-de-da. Esther Fuller? One Moment, please. Is Fuller an expiration? MRS. SCOTT, MRS. SCOTT PLEASE. He whispers he's gonna carry Luke upstairs and he does and I'm running around kicking stuff under the couch and spraying myself with Cachet. Hi, Scotty, outside call asking about Esther Fuller, expired today at one. So he comes back and we kiss and start sinking onto the couch again. That fuckin Luke had wet the couch. . . .

The Red Light started blinking and the buzzer roared. Twelve answered the Code Blue line in seconds. CODE BLUE, ICU CODE BLUE, ICU. CODE BLUE, ICU, CODE BLUE ICU. She rang the nursing office, emergency and respiratory therapy to be sure they had all responded. Five passed her the Code Blue book and she noted the time and location.

> Bet you that was Dr. Strand's patient. Wouldn't have him do open heart surgery on me . . . he'd be diddling the scrub nurse with one hand. . . . Sos he went back and got Luke and brought him back downstairs to the couch and carried me up to bed. Sorry, Dr. Norris is in surgery. May I take a message? Can you believe it, that kid had wet the bed too? Sorry, Out-Patient Billing is closed. No, Sir, I don't know what you can do with your bill. Got a few ideas though, asshole. Right, Nine, you got it . . . the floor was dry. Wish I had a rug. Hey, Five, what's the matter, am I shocking you?

Five had stopped filing, just sat there, pale. "If I'm shocked it's because you girls talk so much and ignore your job."

"Who, there! Dig it, Elaine! Check out my board. One free cord. Next to Thelma I'm the fastest gun in the west. Service with a smile and I could tell you my whole life story and not miss one light. Yes, Dr. Strand. Thank you. NOW he wants ICU."

"It's true, Five," Elaine said, "And you AMs talk just as much as the PMs do."

"Actually they don't, they write more," Nine said, "But PMs get more calls and they have one less operator."

"Five, go home, you look terrible."

"There's still 15 minutes."

"I'll sign you out," Elaine said.

"I prefer to stay."

• • • •

Thump thump at three o'clock when Ten and Eleven punched in at the time clock. Their laughter echoed, their wooden clogs clattered down the halls. They had been shopping, bags rustled and thudded on the counter. Eleven passed around a bag of chocolate chip cookies but the AMs said no thanks. The little room was crowded and raucously loud. The operators bent into their boards, fingers in their free ears.

"You both look adorable in those hair-dos," Six said. Five muttered to herself. Six was forever complimenting them on their hair . . . either Afros or, like now, those beaded corn rows, clacking away. As if they needed extra noise around here. How they ever kept all their men straight. Nine and Elaine didn't talk about their private life. Something to hide, probably. Twelve was passing around pictures of her kids. Poor green-eyed frizzy haired mulattos . . . and she's bringing them up Catholic. Five felt dizzy.

"Elaine, Darcy's so sick with strep throat. Any chance I could leave early tonight?" Ten asked.

"It's up to PMs . . . it's been busy."

"We can handle it. Things slow down at seven. Ok with you two?" Eleven asked. Sure.

The AMs looked at each other. Phoebe would never have stood for this. EA time, sick time, trading days and hours. Three and Four hadn't had one sick day in four years. They called Ten "Typhoid Mary" she was sick so much.

• • • •

Instead of getting quiet, when Elaine and the AMs left, the volume in the little room increased. Ten and Eleven had to tell about their days off and Twelve hadn't finished about Mario. Breakfast in bed! They turned the radio on to low and Ten opened the vent to the sound of cars.

"Phew! Fun*ky*! I swear those women smell like horny lady wrestlers."

Several respiratory therapists came in, to leave the evening list and say hello. Ray, the security guard came in to pick up his gun, asked if they wanted anything on his next round. The PMs didn't take regular breaks, just hurried to the bathroom and took turns making coffee runs.

"Listen to this . . . Dr. Parnell wants us to beep him at 8:30, at 10 and at midnight."

"Wonder who he's trying to impress . . ."

"Or what he wants to get out of. What if we have to beep him for real?"

"ICU wants Dr. Strand. Is he in the house?"

"No . . . he just went out on beeper."

"Far out. Ante up." Each operator put a dollar bill on the board.

"Nine, you realize this idea of yours has boosted PBX efficiency 100%?" They raced to answer the incoming lights, got off each call in record time. Ten got Dr. Strand's call, winning the four dollars.

"Hi, there, Dr. Strand, . . ." she drawled ". . . You always respond so quickly . . ." Laugh. "ICU wants you . . . bye, now . . . He says 'not necessarily' . . . whoee he's a fox."

"Twelve, Mario coming tonight?"

"Not a chance. It's whatserfaces birthday."

"That's the bummer with married men. And Sundays."

"And holidays."

"I don't know . . . has its good points. I went with a fireman once and that was great . . . three days on and three days off."

"Yeah, but what if you're on when he's off?"

"Story of my life. . . ."

The board buzzed like cicadas. It was too busy to talk. Oscar, the Filipino orderly went out to get them Chinese food. He brought it, with a usual kiss on the forehead for Eleven. "Come on, Mama, when you gonna try a flip dude?"

Ten insisted on paying for the dinner, because she was leaving so early.

"That makes no sense at all. We're the ones staying and getting paid."

"I know, but I hate to leave when it's so busy."

"It is busy. Damn."

"Lot of codes, too."

"Notice how those old men in CCU call down all night asking what time is it?"

"Don't you wonder if they're the ones who die later on?"

"Seems like the old women all want something, a cookie or a pillow . . ."

"What does PBX mean?"

"Hell if I know."

"Poor Pay, Boring and Exasperating?"

"Where is Cilantro? I've been paging her for hours . . .'

"Probably somewhere with that new porter."

"Isn't he fine? Big old eyes and nice little ass."

"Can you hear sirens?"

"I did, on a call."

"That fireman, Roman, in 3810, just called down to ask where the fire was. They think we know everything."

"I like it when they want to be connected to the bed by the window."

● ● ● ●

They smoked, sharing a TAB can for an ashtray. The calls were slowing down and they were tired.

"You're so sad, Twelve. That's no way to go *in* to a love affair."

"Well, I'm in and it's gonna be sad. He's Catholic and won't leave her or his kid. And face it, I've got four half-black dago kids and no child support. I know I'm going to love him bad, Nine . . ."

"You sure it's . . ."

"It's worth it."

Eleven and Nine got the lights. Twelve stared, singing "Misty Blue," doodling in red on her expiration list.

• • • •

Monday. Elaine called Nine in to have lunch with her before going to work. Check this out . . . A two page letter to Mr. Hindeman, the administrator. Anonymous, of course. A detailed, nasty letter telling how bad the switchboard was since they had hired an inexperienced supervisor . . . with no seniority. Unqualified operators were being hired, solely on the basis of their race. Expectations were nil. High absenteeism, poor efficiency, bad manners, incorrect language. Calls were not answered promptly, the letter ended, endangering lives.

"How bad are PMs, anyway?"

"Not that bad. Sure they eat and smoke, because they don't take breaks. If there's absenteeism it's because it's the young ones who work weekends and pms . . . they're the ones with children and boyfriends. They all work really well together. The just seem rowdy to the AMs. I'm assuming it's an AM who wrote the letter. Five?"

"Who knows . . . Three and Four are as bad. Or Two, or Seven, she's such a pile of resentment. I don't give a damn who did it. It just hurts."

"What did Hindeman say?"

"He was fine. He's impressed with how I'm doing on moves and changes, installations. Phoebe never could figure those out. No, he's behind me."

"Are you going to post it?"

"Hell, no."

• • • •

No one mentioned the letter and for days everyone was so pleasant Nine began to wonder if it had been an operator who wrote it. The following Monday Mr. Hindeman came into the PBX room. For the first time ever, Thelma said.

"Cozy little place you have here, girls. Doing a swell job, swell job. I've come to take your boss away from all this. Ready for lunch, beautiful?"

"I'm ready!" Elaine waved good-bye.

"I'll bet she's ready." "*Her* job is safe!" "How to succeed in business . . ." were some of the notes.

Elaine came back into the room. "I forgot to post next month's schedule. There are some changes. Bye!"

The AMs never looked at the schedule. Their hours and days hadn't varied for years. It was the PMs, with their turnover and never ending problems, that changed from month to month.

Thelma looked at the schedule when she got up to go to lunch. She rolled her eyes up at Nine, but said nothing as she went out the door. As soon as she was gone Three scurried to the bulletin board.

"Oh, my God. My days off are moved to Mondays and Tuesdays. That's not right. It's not right at all." She tottered back and forth in front of the schedule, sweating. "Four, you still have your days off, but your hours are 10 to 6."

Three sat down at the board and Four went to look at it. "Ten to Six! I can't ride busses that time of night. I won't get home until 7! Five, you're the same, so's Two."

"How about Nine?" Five asked. "Surely her friend has taken her off week-ends."

"Nine is just the same. Twelve has . . ."

"Don't tell me! What's happening?" Twelve arrived to find the room in an uproar, the board bleeping angrily. "All *Right*! Sundays off! I can party all Saturday night and go to Jesse's Little League games! I'll even go to Mass!"

Elaine got back from lunch just before the change of shift. "Smells like gin," Four wrote. "AND enough Arpege," Five added.

"Has everyone seen the new schedule?" They nodded. "We'll try it for a month . . . see how it works."

"Elaine I worked 18 years to get Saturdays off." Three said. "And Seven needs to work four days, not just week-ends and on call."

"The schedule is temporary but it is final. Do all you ladies understand?"

Three and Four were both crying. Three unplugged and left the room. Five scurried quickly over to Three's chair, so that she could page. Elaine moved to stand behind her. She was wearing high heels and her hair was coiled high on her head. She looked regal and beautiful.

'No, Five," she said softly, "Nine will page."

"I will page until three o'clock. I have been here for 23 years and I have the seniority for paging."

"Come over to my desk." Five refused to sit down, stood, trembling, while Elaine put a cassette into her little recorder and turned it on. Five looked bewildered, did not recognize her own voice on the tape, tinny and garbled.

"Why, Five that's you!" Thelma said, pleasantly.

"Me?"

"Sounds like shit, doesn't it? You are not to page, ever again."

• • • •

Ten and Eleven arrived, to join the AMs around the schedule. Elaine plugged in at the board to help Twelve and Nine. No one noticed Five leave.

"I can't believe it! 10 to 6 on weekends! Wonderful! Eleven you got . . ."

"Hush! Let me look! Sweet Jesus! Fridays and Saturdays off! Elaine, you're beautiful . . . Thank you."

"Hey, you guys, come help with this board."

Thump. Who's that? The time clock punched, but they were all there.

"That's Five," Elaine said. "Leaving 10 minutes late."

"Our Five? Late? Jesus Mary Joseph!"

Five could hear their laughter as the elevator doors closed. She had missed the 3:15 pm bus.

WOODSHEDDING IN THE PERFORMING ARTS

Joan Schirle

A bumper sticker on the back of our company van reads, *Where in heaven is Blue Lake?* Where, indeed? Up the coast from San Francisco about six hours, through the redwoods, turn right just past Arcata on the way to Willow Creek and Redding, then just follow the signs. The Dell'Arte School of Mime

& Comedy is the biggest building in this little town, founded by the French as a pack-train station between the coast and the gold fields of the Trinity Alps. Our building, a former Oddfellows lodge erected in 1912, also houses the Dell'Arte Players Company, the only profesional theatre company in California north of Marin County.

Blue Lake is a town of less than a thousand people; lumber trucks move through it on their way to the local mills, the largest industry in Humboldt County. The Mad River is a five-minute walk from town; the Pacific Ocean is ten minutes by car. It is an area in transition, an area whose dwindling resources have sent the county into a downward economic spiral, but whose abundance of water, space and energy have caused the jealous eyes of the lower half of the state to focus squarely upon us.

An unlikely spot for a professional theatre company and school, you say. Yet how many times have you heard actors, designers, musicians say, "I'd give anything to live in the country. I just don't know how I'd make a living there." For some of us here, that desire became so important, so necessary, that we have given up a great deal to be able to make our living doing the kind of work we want to do in the area we want to do it in.

I read a comment by some famous mountain climber who was asked why he climbed mountains. He said, "because there are some things in the face of which one cannot be cynical." I *need* to live in the country at this time in my life; I am by nature cynical, tending toward the negative and pessimistic—it's hard for me even to read the paper. Since I don't admire this quality in myself—it's non-constructive—I must put myself into an environment where I have a fighting chance to remain positive.

Not having to lock my front door or my car helps; being able to walk alone in the streets of my town at night helps; the view of the mountains from my window helps. Doing comedy helps; Christopher Fry called comedy an escape—not from truth, but from despair. I believe it's necessary for each person to find what helps them to be sane, as long as it doesn't involve forgetting, or obliterating, or obliviating for too long, because that's not sanity, that's just being neutralized.

In one of the plays I wrote, BITTERSWEET BLUES, and then later in a play by the whole company called INTRIGUE AT AH-PAH, I was able to create a character that constructively utilized my personal doldrums. She was Scar Tissue, lady detective, a hard bitten cynic if there ever was one, but audiences really took to her. One of my ways of trying to stay sane is to take my problems, my hurts, my negative feelings, and turn them into something creative: a character, a story, a song.

When Carlo Mazzone-Clementi founded The Dell'Arte School of Mime & Comedy in Blue Lake, it was for two reasons: to provide students with an environment free from urban distractions, and to give them the opportunity to learn nature's lessons about performing. In nature, we observe the primary source of a performer's drive and energy. It's that same "force that through the green fuse drives the flower," *that* pushing, driving force that can manifest in the delicacy of a flower or the strength of a redwood. It's the same force that is deeply feared by Western man, the one he keeps trying to put the lid on, to control. There's more than greed involved in the incessant drive to pave over everything, to dam, to channel, to suppress, to make "nice"—there is fear, a kind of "naturophobia"—nature is too awesome, too energetic, too sexual, to *natural*. Translate this fear into an attitude toward the performing arts, and we see a lot of paved roads and milled lumber onstage in the form of safe, easy to traverse ideas whose structures are up to code.

We, in Humboldt County, are at an interface between nature and civilization, a transitional phase which has already been passed by in most other counties of this state. The human advancement here is further along than in some of the mountain counties, and less far along than in Mendocino or Sonoma, for instance. So it's an interesting place to be right now. The future of nature on this planet, the future of our relationship to it, to our resources, is, we feel, being decided in microcosm right here. We are creating an artform which springs directly from this microcosm, this moment.

Some friends in the city say "How can you stay up there, away from all the *real* problems?" Is whether or not the city will have a water supply in 10 years a real problem? When what is sprayed onto trees here makes its way into the food chain and ends up in urban stomachs, is that not a real problem? It's a kind of reverse provincialism—city folks can't see that what's happening up here at this interface is even now having a great impact on the quality of their lives. It all seems like some idyll to them, some fabulous retreat. Yet they're the ones living in the idyll. With the exception of one or two major urban centers, we live in an incredibly conservative state. That's easy to forget or overlook in the city, as is the fact that huge areas of this country are being "re-colonized," in order to exploit their resources. It's not any easier living here, just different.

One of the hardest things is even local people saying, "Well, if you're any good, what are you doing *here*? It's hard not to have a large community of professional performing artists; hard to deal with the lack of understanding of why we have chosen to work here. And there's the usual problems that all theatres face now—lack of money, facilities, support.

On blacker days, I wonder why I keep doing it, but I also know that I can't *not* do it. I'm an actress; I want to work. I also want my work to relate to my deepest concerns. Milan Kundera, the Czech novelist, said, "The struggle of man against power is the struggle against forgetting." So much is forgotten daily, air-brushed from history, from a people's memory: indigenous cultures, other species, events, ideas with low ratings. By creating plays, by trying to give form to our observations and experiences of our lives and times, we can remember who we really are and what our true needs and wants are; and those who see our work are helped not to forget. When we create original plays, we speak for our generation, our time, our situation, not because we want to be spokespersons, but because our work comes out of our experience, much of which has been shared by millions of people.

So many of my friends feel that who they are, the personages they developed and matured during the sixties and early seventies, are obsolete. They feel betrayed, behind and out of step with those who continued to build conformist lives; they feel foolish. Now they are running to catch up, letting careers provide an explanation for their existence. And they feel economically behind, as do I. I don't like being poor; I don't like the anxious feeling of wondering how I'll spend my old age, and the older I get the more I question my choices. But, as my friend and colleague Don Forrest remarked, "We are presently deliberately downwardly mobile in order to maintain our independence in our work." Nevertheless, our present company, created in 1977, was founded on the commitment to living wages for performers. None of us has had to work a second job, and we are now starting to reap some rewards from several years of work in the form of large audiences, critical support, and interesting offers based on respect for the quality of our work. This makes me happy; it makes me feel the methods of working together we have evolved and the influence of our surroundings have been fruitful, that we are evolving an indigenous artform here.

Still, let's face it, our company is made up of "hybrids," or "urban transplants." None of us was born with leaves in our mouths, suckling at the breast of mother earth. I grew up in San Jose, the quintessential suburb, and the rest of our backgrounds are urban or suburban; Detroit, Seattle, Paris, France, Washington, D.C., the Bronx. We had all logged many years of training and working in our profession before coming to Humboldt. We all came at various times because we wanted a change in our lives, a different environment, different values, different rhythms, different influences. We wanted to keep the best of what we brought from our own backgrounds and use that to express our ideas, concerns and feelings about the time and place we live in.

We have been greatly influenced by certain traditional popular theatre forms; commedia, vaudeville, silent film comedy—as well as martial arts, dance, "I LOVE LUCY," The Three Stooges, and other TV comedy of our childhoods. We try to stay open to other influences, too. When we came back from performances in Venice, Italy, we were full of European companies' ideas about mask and spectacle. These were incorporated into our production of WHITEMAN MEETS BIGFOOT.

The work of our company was originally inspired by the commedia dell'arte, which literally means, "the art of the professional performer," and is a form developed in Italy which makes use of improvisation, broad physical gestures, masks, topical references, and stock characters. We have since explored other forms, however, since all our ideas don't always fit the commedia style, and it doesn't lend itself to great subtlety. We have kept the commedia emphasis on physical expression, spontaneity, relating to the audience, and the respect for the performer as the most important part of the production. But we have also explored other acting and production styles: melodrama, cartoon, realism, even conceiving certain pieces cinematographically.

We presently write most of our own work. In each play, by stressing specific influences, we try to create a complete world. For example, we steeped ourselves in Raymond Chandler, Dashiell Hammett and Humphrey Bogart for INTRIGUE AT AH-PAH, a play about the decline of the Northcoast salmon and the future of California waters. With BIGFOOT, based on the comicbook by R. Crumb, we wanted to create the graphic world of Crumb onstage, from characters to music, but our research took us to primate centers, gorilla cages, and books on evolution and anthropology.

For our latest work, PERFORMANCE ANXIETY, we were greatly influenced by the content, though not the style, of Milan Kundera's work— Kundera has written very movingly and articulately about male/female relationships from a man's point of view and, since the play was commissioned by the men's counseling center of a local clinic, this is exactly the point of view we were looking for. The play is about men's responsibility in birth control. Yet, when we discovered that the barriers to this responsibility and involvement include ignorance, taboo, embarrassment, carelessness, and anxieties stemming from age-old myths, we knew it wasn't going to be a simple story. But that's part of the excitement of our working process, which involves a constant interplay of creation and investigation, and is always full of surprises when we start to get beneath the surface of the material.

Our combination of professionalism, attention to craft, multi-layered exploration of our subject, and non-urban viewpoint, has given us a unique theatrical style which seems to have a wide audience appeal. We have helped to develop this wide appeal by touring and playing extended engagements in larger urban areas. This is for several reasons: first, financial—we cannot support ourselves completely in an area where the total population is only 150,000; second, we also want the contact, the exposure, the influences of other professional theatre workers, to keep our own standards of craft high. It is easy to become self-satisfied and to find standards gradually lowering when artists remain too long in the country; third, we have to tour because we do not yet have our own resident theatre space. We became a touring company in the beginning for this reason, no other, and it has provided an interesting, if erratic, livelihood.

We are now, however, ready to build our own space in our own community, and are hoping to complete an outdoor ampitheatre within three years. Our school currently draws approximately 30 students each year from all over the world. We are trying to expand our community base, as well as take our work to European and East Coast audiences. We continue to unify around our work and our ideals, because we like working with each other, and because if we didn't, it would all fall apart. "If you go down in the woods today, you'd better not go alone . . ."

MIDWIFERY: Who's got the power?

Elizabeth Davis

I am a midwife. I love my work and feel I'm practicing a time honored and marvelous art. One that has re-emerged as we've come to question "over-tech," seeking instead some simplicity and intimacy in life's transitions. Historically, midwifery was *the way* of birth until the rise of industrialization and centralization, wherein lie the roots of Modern Medicine. As recently as 1930, most babies in the USA were born at home.

Many of you have probably had some exposure to home birth (your own parents were probably born at home). Or perhaps you've heard stories from friends, exhilarated reports. And maybe you've met up with a few of the original pioneers of the current movement—those unique and stubborn individuals who said "hell no" etc. and stayed home, unattended, for their first child's birth. Many of these mothers then took it upon themselves to obtain information, equipment and skills sufficient for being that special person, the midwife, that they would have loved to have had at their own birthing. From first hand experience, these women learned what it takes to facilitate passage for mom and dad and babe. Passage that is not only physically as comfortable as possible, but emotionally and spiritually harmonious, potent.

And then there are others, like myself, who are survivors of the hospital experience. Unable to find a midwife during my pregnancy, and ultimately unable to analyze seeming abnormalities at the onset of my labor, I went on to the hospital. I'll never forget that moment of decision either—with my labor underway a month early, a two hour drive from our farm to the nearest hospital, snow outside and an old pickup truck for transport! Once ensconsed, it was red tape and a big battle for the most basic rights and courtesies, let alone aesthetics or love. Years later, I summed up the experience to a friend, "like being in jail, on psychedelic drugs, and grieving."

That was ten years ago, and thanks to the midwives and those parents smart enough to use them, times have changed. I myself had a second child at home a few years after my son's birth, in the days when the grass roots finally started pushing upwards. Like most other self-taught midwives, I complemented my own experience with extensive reading and began to teach childbirth preparation, and soon was attending births regularly as a labor coach. When I came to San Francisco in 1977 I joined a midwifery study group comprised of a doctor, obstetrical nurse, and several practicing midwives. With some on-the-spot labor coaching experience in the hospitals and increasing attendance at home births, all my theory came to life (and into perspective) and soon I began to practice on my own. And thanks to various local progressives in the hospitals, backup for emergencies has been available.

I mention all this to show how my eclectic training has been well rounded, and my skills and perspective matured and sharpened by coming into practice gradually. Contrary to popular media hype (and more on this in a moment), most midwives take their work and responsibility very seriously, and are in no rush to take on life-and-death karma until good and ready, lest they jeopardize their clients and their own family life. Most midwives place great importance on doing thorough prenatal care, with plenty of work on fears and psycho-peculiarities, in order to anticipate and ward off potential complications and emergencies. Nutritional counseling, sibling preparation and exercise routines are also part of the package. In a nutshell—holistic *health* care, family centered.

At present, there are 800 midwives practicing in California. Last year 15,000 parents chose to use their services. But these midwives are unlicensed, and their work illegal under the statute of "practicing medicine without a license." Until recently there was little prosecution. But it looks like the days of tolerance by state and medical systems are at an end. Any midwife these days who's unfortunate enough to have a fetal or maternal death will almost certainly be pressed with first degree murder charges. In all such recent cases (and there have been several in the last few years), it was determined by expert testimony that the death would have occurred even if in the hospital. Thus murder charges have always been dropped, but the personal harassment and legal costs are something else again. Recently, for the first time, a midwife was *convicted* of practicing medicine without a license. Her conviction was based on the testimony of several pleased and willing families that she had assisted, families who thought they were doing her a favor by attesting to her thorough, medically grounded and competent care. She (a mother of three young children) was sentenced to a month in jail and three years probation—which puts an end to her practice, her livelihood, and the availability of her services to those in her area who want home birth.

Midwives have rallied in support of one another, . . . raised funds, and formed defense organizations. But, politically speaking, well . . . here's where the paradox comes in. Because, basically, midwives are anarchistic (or apolitical, at least) and have a terrible time among themselves evolving the structure, the standards of practice which might make them politically viable. The reasons for this are several: 1) although modern midwifery includes technology and standard treatment in emergencies, birth remains a unique process; never twice the same so that, literally, what works (and is safe) in one situation might not be right in another; 2) midwives themselves have such diverse skills and expertise that, whereas one might feel comfortable in assisting, say, a breech delivery at home, the next wouldn't touch it with a ten foot pole! There are, however, certain bottom-line protocols for safe home practice. These basic tenets, plus an educational proposal, were formulated into a bill for licensing which came before the state legislature this spring. The bill was defeated (for the third year in a row) which is not surprising, considering the volume of anti-homebirth and anti-midwivery literature that each legislator and every member of the California Medical Association received in the mail prior to the vote. The midwives, to put it bluntly, have not had the money or the sophistication for such tactics, but there are some who are considering this direction.

Then there are others, myself included, who fear that if midwifery is legislated it will lose its power as an art form and become just another government regulated business. In this respect, one thing is certain—although a prescribed set of skills and information may make a midwife technically competent, this will not insure her sensitivity, intuitiveness, or courage to assert her intelligence in innovation, *especially* if the schooling is to be designed and administered by the medical establishment! Take, for example, the hypothetical case of a woman with hypertension (high blood pressure). She would immediately be screened out as a candidate for home birth, by current medical standards. However, most midwives know that a combination of herbal remedy, dietary changes, and recommendations for either rest or exercise (the exact combination depending on the degree of elevation in pressure, the woman's lifestyle and personality) will not only stabilize blood pressure, but will lower it significantly, in most cases. This is but one example of the art of midwifery. Midwives' journals and medical records are full of alternatives to standard medical treatments which involve drugs or technological devices.

So how do we keep the art of midwifery alive and free? I think that it all depends on the informed consumer. (Power to the people!) If prospective parents are hip and well educated regarding their choices of birth, then hopefully they won't be duped. Beyond the media flashes, there are some excellent books that explain the complications of labor and pregnancy, management of these problems at home, rights in the hospital, aesthetic possibilities in *any* situation, care of the newborn, and the various psychological aspects that might affect different phases of the experience. The informed consumer is not easily deceived! But misinformation is a problem, too. There is an abundance of birth-related promotional hype for hospital services that deserves to be investigated and exposed. For example, the alternative birth center promise of a "home like" atmosphere (whatever that is!) frequently does not come true, due to understaffing, strict regulatons for transfer-out in case of minor complications, or simple overbooking of facilities. If parents know and understand these things, they they can take responsibility for choosing the setting and the assistants with the necessary skills, experience and character to suit them.

I recently wrote a book (*A Guide to Midwifery, Heart and Hands*) which most of the folks I assist have read. And believe me, they challenge me, talk back, and propose alternatives to my suggestions. This is the way I want it because it keeps me from getting set in my ways, keeps me in *service* instead of control. So if you've a mind to do so, get involved when your turn comes round, and share your experience with others. Speak out! If parents insist on being treated as individuals, and we midwives are there to listen, then the power of birthing cannot be usurped.

Ed Buryn

How to Play Hardball

The Great Bastard never said good morning. He never said please, never said thanks. The Great Bastard never drank with anyone and never ate with anyone. But, everybody said that, deep down, he was Santa Claus. He never showed it.

One day, in the company lunch room, I overheard two guys plotting to get the Bastard. Nothing new. I called his secretary. When they came back to the office, their desks were gone, Security waiting. I got back to work.

Without warning, his face expressionless, the Great Bastard strode into my office, took a red ball from his pocket and tossed it to me. I jumped up and caught it. We played catch, expanding our distance. We were flawless. We threw harder, faster, grunting with the effort. Neither of us came close to making an error. Then, abruptly, it was over. The Bastard put the ball in his pocket and walked away.

From that day on, he never said good morning. He never said please. And he never said thanks.

—John Lowry

POLITICS

his morning's mail:
a credit-card bill from Exxon
& a draft notice
    —Pete Beckwith

POLITICS 215

Joseph Carey

FRANK'S WAR: Vietnam 12 years after
Frank B. Kiernan, talking

I was in Vietnam 11 months, 3 weeks, and 1 day. I went there at the beginning of 1969 and came home at the end of '69. I was in the Army Cavalry. I was drafted when I was 20, shortly after quitting college, while working for United Parcel in Connecticut. Out of my infantry unit in basic training, they made 78 people regular infantry and 4 people medics. I was one of the medics. I was in the 1st Squadron 1st Cavalry which is tanks and armored cars.

I arrived in Saigon and then, about a week later, was sent to a Base Camp 27 miles south of Tam Ky. I was there about 48 hours before realizing that I had made a mistake, that I never should have gone to Vietnam. I realized that the people we were presumably there to help didn't want us there. I felt a lot of hostility from the Vietnamese people. The fact that they didn't want us there made me feel like I shouldn't be there. If I had been more aware of the real situation before going, I probably would have done almost anything not to go there. At the point that I did go there, I thought I was doing the right thing. My family was behind me in the respect that they had left it up to me as to what I should do. Being uninformed and patriotic, I accepted the Draft and went to Vietnam.

The main thing I realized when I arrived there, after getting into the field, was that I couldn't think about what was happening in front of me. Because if you thought about what was going on you would really lose your mind. So instead of thinking, you reacted. If something happened, and someone's guts got blown out in front of you, you didn't think about it. You just reacted and did what you could for the person.

I don't know why they made me a medic. Being drafted, I had no say on what my MOS would be. I had never held a gun in my hand until I went into the army. But, they didn't know that. I was glad I was a medic, though, because I think the worst thing you can do in the world is take a human life. The positive side of having been in Vietnam for me was that I saved a lot of human lives, both North Vietnamese and Americans. The lifers respected me because anytime someone called that was hurt I went to their aid. After 16 months I made sergeant because on one occasion there were 4 medics above me who were destroyed in one action and I was the next guy in line. So they made me a sergeant.

When I was there, there were a couple of different groups of people. There were lifer-sergeants who were gung-ho, charge the hill types. A lot of them ended up getting shot in the back with M-16s by their own men. There were intelligent officers, who had graduated from West Point, who would call into the colonel and lie, "we've already been to this hill." They knew there was no sense going up there and getting our asses kicked for nothing. They were at the point where they would fake a report saying we went there when we really didn't go there. Whereas a sergeant, the lifer, the E7, who is in for twenty years, would want to charge the hill. And those were the people that got the three warnings. They got the smoke grenade, that's the first warning. The second was the CS gas, and the third time he said charge he got fragged to death. They fragged him. I saw that happen three times.

The whole body count trip was pretty insane. They didn't just count dead V.C. soldiers. They would count mama-sans and baby-sans. They would count 75 year old papa-sans. If we went through a village where there was a mine that was triggered 200 meters away, some GIs, and I can see why they did it, would shoot innocent people because some of our people got killed. They would shoot people that had nothing to do with that mine. I saw that once or twice. I was in Mai Lai a year after it happened. Mai Lai was the only village that I was in where I saw booby traps hanging in the trees. It was the only place where we were supposedly entering a safe village where I saw booby traps. So, we stopped and dismantled them.

The Marines and the Airborne fought the real heavy war. They probably saw combat every other day, every third day, whereas I saw it once a week. One time I didn't sleep for four days, under the influence of no drugs, out of pure fright. But, the Marines and the Airborne people were in a lot more shit than I was ever in. But, I was in *enough* to scare the shit out of me.

I saw *Apocalypse Now*, I saw *The Deerhunter*, I was never in anything like *Apocalypse Now*. I'm not sure what Coppola was trying to do with that movie. I think he was trying to make an excellent movie. He tried to do it too hard and ended up botching the whole thing. I did, though, see some insanity similar to that. I was in Vietnam about a month when this colonel, who was in charge of our battalion, which had 9 tanks and 27 armored cars, took us out into the field to take pictures to send home to his wife. One of our tanks hit a mine and three people got killed. The only reason we were there was for him to take pictures. When that tank hit the mine a guy that had been there 11 months went down on his machine gun and aimed it right at the colonel's chopper. It took four people to subdue him before he got off of his gun. The chopper went from 100 feet to as high as it could go, as fast as it could go. That colonel was taken away and I never saw him again. That was the most insane thing I saw. I thought *The Deerhunter* was a much better movie, although I was never attacked by 50 people at one time. I was there 11 months 3 weeks and 1 day and was in combat situations 48 of those days.

I spent nine months in the field. The typical infantry grunt spent about 11 months, medics spent about nine. And after nine months I told the guy that I was too nervous to spend any more time in the field and that I was a wreck. So for the last two and one half months I was in Base Camp where we just got rocketed instead of attacked. I spent about 2½ months at Base Camp and then I was shipped home. I was in the army 21 months. I got out early because I was drafted and had been in combat situations.

I was pretty lucky. I was wounded very slightly twice. So that was no big thing. When I came home I made a point to forget about what happened over there. I don't talk about it too much. Sometimes, when I get very high and I might be around someone else who has been there, I might talk about it but I don't dwell on it. When I was there I didn't think about it, I reacted, because if you thought about it while you were there, you would go nuts. When I got home, I forgot about that year of my life. That's basically how I dealt with it. And I really don't have nightmares about it. I feel I am lucky because I'm sure there's a lot of people, besides being physically maimed, that are emotionally scarred. And might be emotionally scarred for the rest of their lives.

A couple of things persist from my Vietnam experience though. For example, I cannot sleep on my stomach. Because when you are over there you always sleep on your back. You always sleep with a gun in your hand because that way if there is a zapper attack you have a gun and you're ready. I always slept outside on my back, even in the rain. Because the guys that slept in tanks or other enclosures were prime targets for grenades and mortars. Outside, on my back, I could hear things and be much more ready. Even though I was a medic, if someone was trying to kill me, that was the only time I shot back. Very seldom did I shoot back because almost all the time we were shot at first and I was called upon to help someone. But the basic thing, since I came back, is I always sleep on my back, I don't sleep on my stomach. Because over there if you slept on your stomach you could wake up with a knife in your back.

I was very glad I was a medic even though I had no choice about being made a medic. I was sometimes called a gook lover because I helped wounded Vietnamese, but I was also respected because I always helped anyone that called upon me. I was awarded a few medals while I was there. Besides the Purple Heart, I was awarded a Bronze Star and a Silver Star.

Another thing that ocassionally crops up in my mind is a memory of a dying soldier. You know when someone's going to die. You can look in their eyes, you know they're not going to live, and you just tell them, "you're lucky you're going to Japan." But you just have to set them down in the helicopter and help the next person that you know has a chance of living. Sometimes I didn't think I should do it. I don't think anyone should do it. Deep down in your subconscious, you think maybe you could have saved that guy's life. But there's no way you could have. As hard as it sounds, when you look at their eyes, they glaze over. That's sort of the point of going into shock. Their eyes glaze over and you know that there's really no basic value in spending time with them when there's three or four other people that definitely have a good possibility of living. I'm not saying like if there was any chance at all that this person would live that I would leave them. But after being there a month and seeing a few people you know when someone's not going to live more than five minutes. After Vietnam I had no desire to pursue my medical experience. I did the exact opposite. I never wanted to have the medical responsibility for someone's life in my hands again.

As soon as I was discharged I called my parents who lived in Connecticut and told them that I was okay. That I was going to stay with some friends in Oakland for two weeks and then go to Connecticut. My friends picked me up at the San Francisco Airport and gave me some LSD. I took a tab and started peaking with it on the San Mateo bridge. I stayed in Oakland for about two weeks and then I went home. It was one of the most sentimental times in my life. My parents and whole family came to the airport to meet me. It was the first time I saw my father cry. It was one of the closest feelings I have ever had with my father. That was a very high point in our lives.

When I got back to my parents' home, I lived in Connecticut for about 4 months that summer before I moved back to California. During that time I had four friends that had to go for their physicals to go into the army. I stayed up with them all night and talked to them about how insane it was over there, how the people didn't want us over there. We drank some coffee, got them really hyper, so their blood pressure was extremely high. After talking to me they had to stay two days for their physicals. But they got themselves up for the second day and flunked the physicals. I helped get four of my friends out of the army on high blood pressure that way. Because I saw what was going on over there and realized it was worthless. I would never go to war again unless my own country was attacked. I would leave, I would go anywhere. If they drafted me to go to war in east jockstrap, Alaska, or in Japan, because we were supposedly their allies, I would not go. The only way I would defend our country is if our actual continental United States was attacked. Otherwise, I would not go.

I basically tell people I know that didn't go to Vietnam that I love them and I'm really happy they didn't go. Because I knew immediately that it was the wrong thing for me to do. But I had a choice to make and I was young and I just said, well, I don't know what's really going on so I'm going to see.

Since Vietnam I haven't gotten involved politically, I do know one person that died from Agent Orange. And the government has done nothing about it as far as I am concerned. The government has done nothing. They are either ignoring the fact or they're saying that's bullshit. I know a guy that was 28 years old and he went to the VA Hospital day after day after day and they just shined him on. He died with a wife and two children of that disease. That pisses me off a lot. I mean, there's no doubt in my mind that Agent Orange killed him.

I recently read a book, *Born on the Fourth of July* that was by a guy who became a paraplegic in Vietnam. If something like that happened to me I would probably be very politically involved now. I feel very lucky that I got out with minor injuries. My basic feeling is that unless a multitude of Vietnam veterans get together, which I don't think is happening here, nothing's going to benefit us that much. I think we've gotten screwed. I think a lot of Vietanm vets have gotten screwed. I mean, if I was in a VA Hospital with no legs, I'd be much more involved. Because I've visited people in VA Hospitals where I've seen rats underneath their beds.

I don't read the newspaper an awful lot but as far as I am concerned I think El Salvador would be a similar situation and I think that we have no right to be there. I think we had no right to be in Vietnam. I think it was a big mistake. I think that Reagan or the establishment's going to do in El Salvador what they want to do. I also think there aren't that many Vietnam veterans that are going to try and stop what's going on in El Salvador. I might be wrong on that point, but I still think the government's going to do what they want to do there no matter what all the Vietnam veterans do.

If there was a multitude of Vietnam veterans who protested, then I would join that group. But, I don't think that would happen. If it did happen I would get involved . . . to an extent. I am not going to go out and get killed in the street opposing our position in El Salvador. I would, however, definitely, peaceably demonstrate against our involvement in El Salvador.

The greatest feeling about being in Vietnam was knowing that I definitely saved many peoples' lives. North Vietnamese and Americans. I did the best I could.

Vietnam Haiku

Jittery;
no sign of movement
except the fireflies

No enemy seen;
but I get a good look
at myself

In rear truck
of a long convoy—
the dusty road

Crickets stop chirpping;
I awake
with a start!

—*Ty Hadman*

Brent Richardson

NEIGHBORHOOD CRIME PREVENTION

Ernest Callenbach, talking

About two years ago, when we had just moved into a block on the north side of Berkeley, my wife Christine Leefeldt's Volkswagen was stolen one night. She was desolated, naturally, and reported it to the Berkeley Police Department. In her contacts with the police, she learned that they had a social worker on the staff, a woman, whose job it was to promote block organizations. Because the Berkeley police force had found that the best way to prevent crime is to have people know one another on their blocks. This provides a social pressure against anonymity and crime, which is really the essential thing since no society can ever have enough police officers to watch all the potential criminals. (If they did have that many, then you wouldn't want to live there—it would be a police state.)

Anyhow, this city worker—Virginia Jensen—came around to the house and talked to Christine and explained how we could have a neighborhood organizational meeting and get everybody together in somebody's house and talk about these problems. Besides the theft of our Volkswagen, there was another funning thing happening at the time—kids were joyriding down the blocks, smashing in the windows of foreign cars, especially Swedish cars for some reason. This was very upsetting to the people who owned Volvos.

So we were all a little edgy at the time. Nobody had been burglarized but the situation still had people rather wrought up. When Christine heard about the possibility of organizing the neighborhood, she undertook to be the block organizer and went around and talked to everybody. What the city recommends, and it seems to be a very sound plan, is that somebody on the block take the initiative and go around and actually make a map of where everybody's house is, showing who lives there, and their telephone numbers. If you want to get as extensive as our list, you can also list the names of the teenage children, the little children, the cars, the dogs, everything. So, it can become kind of a directory of the neighborhood. Our street is a very settled neighborhood; very few people move in and out, most people own their houses there.

So we held a meeting to which Virginia Jensen came as well as the local block police officer. He gave us a rundown about burglary prevention and taking care of your car to minimize the likelihood of it being ripped off. We all learned a lot. For instance, car thefts go in epidemics—that week blue Darts and white Volkswagens were being stolen. It's very quirky and personal when you get down to dealing with crime on that kind of level. Then there came a moment when Virginia said, well now, if you are going to do this thing, you have to have a contact person. If you have anything untoward happening it's good to have someone who'll be responsible for reporting it—we'd really like to know about it. Maybe we can do something about it. So who would like to be the coordinator?

Christine, at this point, got elected by unanimous default. She became block captain. And the fact that we were then organized—we had our map and a block captain—meant that we could become an official part of the city's "Neighborhood Watch" program. They gave us red stop sign decals to put on our front doors and whatever doors and windows burglars might find attractive. They say something like *this neighborhood is organized and the neighbors are watching you.*

Not long after that we had a block party. We got an official city permit and closed off the street. The city gave us a barricade for the street. We had a volleyball net set up, a keg of beer, and everybody brought out pot-luck food. It was a wonderful event. We all had a terrific time. For the first time, because of this concerted campaign to get everybody on the map, everybody on our block (which is about 20 houses) literally knew who everybody was. That was a critical development, because if you know who everybody is, it gives you a great feeling of security, of familiarity and shared turf. If somebody is walking around peering into windows and you know that this person doesn't live there, then you know you ought to do something about it. It made us all feel that we could rely on each other a lot more than most neighborhood people feel.

People did express some worry about losing their privacy. It's a thing that obviously concerns a lot of people. But I don't know of any cases on our block where it has become a problem. People have rather easy ways of putting a stop to an invasion of their privacy if it gets out of hand.

We've had two more block parties since. We usually have two a year, one in the fall and one in the early part of the year. And we haven't had any burglars. There was one attempted mugging down at the end of the street by the bus stop where there are a good many strangers. The woman started screaming and three or four people immediately came out of their houses to chase the assailant, while others called 911 and yelled for help. So we considered that this was a good sign of our neighborhood solidarity. In many blocks somebody starts screaming in the street and nobody comes out. They may call the police, or peer out the windows, but they don't actually run out in the street. I personally feel that if some dangerous character entered my house and was about to harm me, if I started screaming out the windows, numerous people would appear on the scene and do something about it. This is not a feeling I've had in most other neighborhoods I've lived in. It gives you a sense of shared turf, a feeling that you and all the neighbors *belong* here, by God. And if anybody comes in and gives somebody a hard time, you're going to do something about it. You don't feel rootless and at the mercy of strangers. And that's a wonderful feeling to have.

Our block group hasn't done anything terribly dramatic. It just started out in response to a crime and it led to these very nice extensions of a local party tradition. But soon I'm going to try to organize a small buying club. We could go down to the coop warehouse every three or four weeks, something like that, and drag back a load of staple stuff. We don't want to buy things that are perishable, but there are a lot of staple things that we all buy all the time in small quantities. I don't remember the precise discount, I think you get either ten or 15 percent off. And once you've bought it, you don't have to worry about it, you've got it. In your basement you could have a carton of toilet paper and only have to buy toilet paper once a year. That would be a really nice thing, in my opinion. So I'm going to try and get that going. It may only involve three or four families. There are, of course, many people around Berkeley who are buyng food together—sometimes on a neighborhood basis, sometimes not. But if it is a neighborhood thing it makes it a lot easier to divide things up. You make your run and then you come back and dole everything out to people, collect the money or however you work that.

On our block everybody seems to feel that it's really nice to know the neighbors. There are no great neighborhood feuds or anything on the block. It's very pleasant—the way town living ought to be and too seldom is. I often hear people expressing envy of our block organization. They say to me, "I only know three or four people in three or four other houses on my street." I say to them, "Why don't you systematically go around and say hello to people?" I don't know whether it's shyness or laziness or fear of rebuff which inhibits them. There was one family, for example, on our block. When Christine went down and said, "We're going to have a block gathering to discuss what we can do to get everybody to know everybody and cut down on crime," this woman said, "Oh well, I'm not interested." But by the time we got the map together she discovered that she was left off the map. Then she hustled up the street and said, "Gee, why aren't we on your map?" Christine said, "Well, you said you weren't interested in the block group so naturally we didn't put you on the map." And she said, "Oh well, of course we want to be on the map." She and her husband came right around for the first party. Maybe they were just organized out; there are some people in Berkeley who belong to so many organizations that the idea of one more is just too much for them to bear.

Our greatest collective problem is that we have a stop sign at the corner which is not very well observed. One of the children on our block was killed before we moved there. It evidently wasn't the driver's fault—the child ran out between two cars. But it's a narrow street and drivers don't always exercise prudent speeds. Various people have contacted the city and asked them, "When are you going to get around to repainting the white stripe at the stop line and maybe move the stop sign up so that drivers can't claim they didn't see it." And nothing happened. We thought we might have to do a direct action number. Get our white paint buckets and go out there one night when there wasn't much traffic and paint it white. If that didn't slow the cars down we contemplated digging a bump. There's nothing like a bump or a little ditch to slow cars down. Bumps are an underused mechanism for the control of automobiles. You only have to drive over a bump like that too fast about twice, and have your teeth jarred a couple of times, and after that you really slow down. So, we were going to make our own little bump there. But the city about that time got its act together and came out and repainted the line, and put the stop sign in a more visible place. It seems to be working better now.

I know of one group of neighbors who did in fact build their own stop sign, at Vine and Edith. Do you know what the city did? The neighbors had put up signs to stop the traffic running east/west. So the city came and took down their homemade red and white painted signs, and turned them around to stop the traffic going *north/south*. Just to sort of exert their *control*; there was no reason to do it, except that they didn't want the neighbors to get too much of a feeling of power. And the stop signs have been there ever since.

I think there may be a new tendency for things like that to happen in connection with cars, especially in neighborhoods where people are taking steps to defend themselves against cars. Under Proposition 13 the city doesn't have the money to repair the streets anyway, so people may just go out and dig a couple of bumps some night, which the city will probably take six months to get around to repair. That will slow traffic down. . . .

Getting a neighborhood organized represents a fair investment of time. Somebody or a few people really have to care about it enough to make it go. Christine must have put in the equivalent of three or four good evenings of work on the Neighborhood Watch Program. Most of it happened in the evening when people were home. She did a lot of going up and down the street knocking on people's doors. To each person you've got to explain the whole thing. What's going on, why it's useful. I suspect that a good time to do it is when there has been a crisis in the neighborhood; there's been a burglary, an accident, a car is stolen, or something like that. So, if you are an aspiring organizer, you could probably do worse than to wait until something of that sort happens in your neighborhood and use that as a kind of organizing tool— a stimulus to get people to realize that they aren't powerless.

American Dreams

George Washington was never America's hero
no matter how many dollar bills they put his picture on
It was Jesse James, Billy the Kid, John Dillinger
set the legends flying
The outlaw is the archetype of glory
the secret hero of suburbia
nobody wants to fuck a cop
but the thighs of America's women grow steamy
as they dream of marauders
the paper pushers dream of violence as they fill out forms
the outlaw's dance with death
buggery and rape, drunken kaleidoscope of suck/fuck gangbang
haunting the secret dreams of the American tracts
THE HERO OF THE AMERICAN ID IS THE MOTORCYCLE OUTLAW
racing his screaming chopper
down the raw-nerved hypocritical spine of the country
unwashed, uncombed, his genitals scummy with yesterday's perversities
the skeleton of his future invoked in present-tense acceptance
riding invisibly obvious beside him NO FEAR OF DEATH!
(a million housewives creaming in their installment plan panties
yearning for luscious domination, the brutal male ramming his life size penis

continued

CRAAACK through the opaque hymen of television boredom, seven shadowy
figures demanding indecent satisfactions, pauseless obscenities, helpless,
ravaged, she's coming like a maniac, but it's all fantasy baby, married
twenty years nor ever sucked a cock and don't forget the young ones
rubbing their beautiful pussies against imagination's ultimate phallus,
god through gangbang, abandonment through I couldn't help it, he stuck
it up me when I wasn't looking, I came without consent)
fascination/obsession, hero of dirty daydreams, VIOLENCE goddamn madman
armed with guns knives chains tire irons brass knuckles stomping boots
KILL THE MOTHERFUCKER he'll do anything smash your teeth fuck you in
the ass kill you rob you violate you TERRORIZE YOU
(oh the righteous indignation of the deodorized
as they yearn to kill niggers, rob banks, cheat the system, smash the
dissident, and dip their whining cocks into the dirty-cunted dirty-eyed
dirty-girls that they fantasize while crawling for the stale delights
of nagging matrimony)
and he sees his dreams made manifest, Mr. America of the mercantile empire,
the hierarchy of ulcers
the tenacious yearning to explode anxiety into violence, the cringing nausea
of repeated compromise, the longing for total commitment

the outlaw has released the fear of fear
inescapably alien, he walks with his death as a garland

there is no illusion of security
dancing a tightrope in the now of forever
the acrobat of disaster juggling his own eternity
gambling his skull against a case of beer
his blood against a definition of honor
THERE'S NOTHING TO LOSE
 on the other side of fear
(and the mortgage, the life insurance, the pension, old age, taxes,
the invisible mechanized rat gnawing the possibilities of man)
there is a glamor in danger
teasing the reverie of a man unfulfilled with himself
there is a magic in the burst tabu
that twinges envy in those who fear their own dreams
the vision of the outlaw
using it all up   playing for keeps
enthralls and appalls the suburban inheritors of the American dream

descendants of
witch burners
slave breeders
Indian killers
treaty breakers
land stealers
sweatshop owners
whorehouse landlords
segregationist ministers
expedient politicians
overkill generals

sons and daughters of the Bill of Rights and the padded expense
account, blind-minded to the petty deceits of historic hypocrisy
waving a con-man's flag
it's your secret brain that shudders at itself
and claims the outlaw as hero
because he does it right out loud
because he wears his own shit like a halo
and lays his life on the line
not for a question of political right and wrong
but for his own definitions
backed by his existence
or his non-existence

and the outlaw is America's hero
secret companion of frightened fantasy thrills in the interchangable
vacuums of dissatisfaction
because too many men have sold out their manhood
trading their visions for time payment lies
and too many women have bartered themselves for a facsimile of love
designed to impress the stranger
while their own hearts grow numb

and the outlaw is America's hero
because he is what he is
and neither begs pardon nor forgiveness nor mercy
and the nation is corroded by guilt

—*Lenore Kandel*

Ed Buryn

TRUST THE LAND
David Prowler

For people to whom real estate is a game, the San Francisco Bay Area is a great board on which to play. The rules are quite simple—buy some property, fix it up or not as you wish, then raise the price tag and find another player to unload the land on. The next player does the same. Obviously, this game cannot go on forever. But although it cannot continue indefinitely, even whopping interest rates haven't dampened the action appreciably. Just yesterday, for example, I heard of a house a few blocks from me which sold one and a half years ago for $125,000 and is now on the market—unimproved—for $400,000.

The reasons why the Bay Area provides a particularly fertile breeding ground for real estate fever are simple and to a degree interrelated. The area has seen a boom in white collar job opportunities, including financial paper pushing in San Francisco and the throwing together of microchips in the Peninsula's "Silicon Valley". A high demand for the available land is created by the sheer pleasantness of the region: the weather is usually pretty good. The hills, ocean, and bay are all attractive. Central city living has been made more attractive by the high fuel costs involved in long suburban commutes. And the area provides a handy haven for foreign investors made antsy by political strife in their home lands.

It all stacks up to a highly profitable game for the skillful wheeler-dealer. As an investment, real estate in this environment has a lot going for it. It is an extremely limited commodity for which there exist not one but two markets: consumers who want merely to enjoy the use of the land, and would-be land barons who want to purchase it to resell at a profit. That's why so many people can gaze out at a landscape and see nothing but "best and highest use", and why fully one out of every nine Californians has a real estate license.

It is a high stakes game, and for every winner there is a loser. The effects of this kind of real estate speculation include displacement, people paying half of their incomes for rent, high land costs, and even arson. Some examples:

The tenants of the International Hotel, some of whom had lived in the same room for twenty years, were physically evicted by the Sheriff of San Francisco as thousands of the tenants' supporters chanted outside on Kearny Street. It took the Sheriff and Police Department all night to clear out the elderly tenants. Today, four years later, where the hotel was is now a big hole on Kearny Street.

Lannear Delaney was a black barber on Haight Street. He'd had his shop on the corner for ten years, since the Redevelopment Agency tore down his old shop. He had built up a pretty good business and his barbershop was virtually a social center for folks of all ages. Then a lawyer bought the building and evicted Lannear. The storefront was vacant for years. Now it's a "Victorian Pub".

The Gartland Hotel was a firetrap and everyone knew it. The building inspectors knew it, the long-term tenants knew it, and the community at large knew it. In fact, a group called "Operation Upgrade" picketed the building to protest its condition. Two weeks later someone poured gasoline down four flights of stairs and burned the hotel down. Fourteen people were killed, twelve were declared missing.

These aren't isolated flukes. The number of people who have to move because of real estate games is enormous. In 1980, San Francisco Municipal Court handled 4500 evictions—which doesn't include those cases where tenants didn't resist, or cases held in Small Claims Court. Although many of these displaced persons move elsewhere in the Bay Area (from San Francisco to Oakland is a common move), many people cannot afford to remain in the region. This process dislocates not only individuals or individual households but entire communities as well. The International Hotel was the last vestige of Manilatown, a once-thriving neighborhood which has been engulfed by highrises. The Western Addition, where Lannear Delaney's barbershop was, used to be quite uptown for San Francisco's black community. Part of it was "Ninonmachi": Japantown. The Japanese were interned in camps during World War Two and returned to find their neighborhood sold out from under them. The Redevelopment Agency tore down the Fillmore to such an extent that now two blocks on Fillmore Street are suitable only for community gardening—and the Agency has given the gardens a thirty day notice to get out. Entire neighborhoods in Oakland were erased when the planners drew in the Bay Area Rapid Transit system and the freeway.

San Francisco used to have a neighborhood called South of Market, but it's hardly a neighborhood anymore. Ben Swig gazed down from his Fairmont Hotel on this community of Irish working families and old sailors and had a vision of a convention center, hotels, restaurants, even a sports arena. So the Redevelopment Agency tore down the homes, corner grocery stores, bars, everything but a church and a Pacific Gas and Electric plant. Their disregard for the natives led to lawsuits which tied up their masterplan for twelve years while the three downtown blocks sat vacant.

Manilatown, at one time San Francisco's Filipino neighborhood, is also gone now. The center of that community was the International Hotel (built in 1850). The Hotel was the home base for the seagoing Filipino men, where they would stay, visit their friends and pick up their mail when they weren't on fishing ships or canning in Alaska. As the men and women who lived in the Hotel aged, they settled in and formed a community of elders—"manongs" in the Filipino Tagalog dialect—relying on each other. Their rents had paid for the Hotel many times over. But when real estate tycoon Walter Shorenstein, and later the mysterious Thai whiskey king Supasit Mahaguna, told the tenants to clear out, nothing they did could prevent the eviction. And they tried *everything*. The tenants sued the owners and they sued back and the tenants lost every time. They put propositions on the ballot and never came close at the polls. The tenants got the City to pass policy statements and it was a waste of time. They wrote laws and got legislators to introduce them and never got more than a couple of votes. The Sheriff even did a stint in jail (not his) rather than evict. On eviction night the tenants and their young supporters carried out paramilitary nonviolence, complete with secret codes, barricades, walkie talkies, and even giant wooden forks to push ladders off the roof. But the eviction was properly signed, sealed, and delivered and the building was emptied and demolished. Manilatown is gone now; soon an office building will sit there.

Think of these communities as the urban equivalents of endangered ecosystems. They contain networks and symbiotic relationships, mythologies and traditions, landmarks and folk art. These neighborhoods have their own names. "Nairobi" in East Palo Alto; "The Iron Triangle" in Richmond; "Dogpatch", "The Tenderloin", and "South o' the Slot" in San Francisco, "West Cypress" in Oakland. When these neighborhoods are gone, they're gone. It's not just households, it's lock, stock and barrel. Stores. Movie theatres. Restaurants and bars. Even the names are forgotten. Dogpatch is now "Lower Potrero Hill", the Tenderloin is now "Central City" or "North of Market", and South o' the Slot has become "The Yerba Buena Center".

This destructive process of speculation/displacement happens to farmers on the fringes of cities as well. Little by little their farms are being bought out from under them. 708,000 acres of Bay Area farmland have gone out of production since 1949, and every year another 19,000 acres—equal to two-thirds the size of San Francisco—shut down. The farmer and the city dweller are both victims of the same process. And they will remain subject to this alienation from the land as long as the earth is viewed as a commodity. In the Bay Area in particular the land is seen as an investment, the same as hog belly futures, krugerrands, and misprinted postage stamps.

This perspective of the land is fairly recent in North America, having arrived by ship from Europe. The Native American Indians had no conception of private "ownership" of the land. A Blackfoot chief, asked to sign a treaty selling land near the northern border of Montana, replied. "Our land is more valuable than your money. It will last forever. It will not even perish by the flames of fire. As long as the sun shines and the waters flow, this land will be here to give life to men and animals. We cannot sell the lives of men and animals; therefore we cannot sell the land. As a present to you we will give you anything we have that you can take with you; but the land, never." Land was seen as community property, for everyone to enjoy and no one to own.

Common ownership of land in North America is a concept as old as human life on the continent. Examples include the oldest pueblos, Boston Common, and Co-op City (America's largest apartment development). If we consider land owned by the government as held for the common good (and under the current administration, it is an open question), then we're talking about a lot of land—the National Park Service alone holds 76 million acres, and that is just one department of a single federal agency.

Another mechanism which a community may use to hold land is the "land trust". It is just a non-profit corporation which can hold, develop, and maintain land for the common good. Land trusts are becoming increasingly popular. They are formed to hold land for a variety of uses. Urban land trusts in Oakland, Richmond, Newark, New York City, and San Francisco, for example, have come together to hold land for community gardening and recreation use. Oakland's Santa Rita Land Trust was started by a group of neighbors for the purpose of acquiring a hillside which was not only an overgrown, trash-filled eyesore, but threatened to slide into their houses when it rained as well. With a little help from their friends (including donated labor from the National Guard), they are converting the land into a community park and playground. A land trust is forming in Berkeley to acquire the largest factory in that City, the 10.5 acre, 24 building Colgate-Palmolive soap factory. They want it to house local co-ops and small businesses when Colgate makes their planned move.

California has its share of rural and agricultural land trusts as well. The Humboldt North Coast Land Trust, in Trinidad, was formed to preserve a scenic stretch of coastline, as was the Big Sur Land Trust. Farmers in Marin County have formed the Marin Agricultural Land Trust in an effort to preserve that county's rich agricultural and grazing land.

Land trusts are the vehicle for community ownership of the land. They can determine the use of the land (the social "best and highest use") and remove the site from the speculative market—in effect, posting a "Not for Sale" sign. The process of forming a land trust and acquiring and maintaining land has its own value. It is wonderful to see neighbors working together for a common purpose and often the community-held land becomes a social focal point for the community. An incident which took place at Oakland's 39th Avenue Land Trust community garden illustrates. Two teenage boys were seen having a tomato fight in the garden. When their parents offered to pay for the damage, the land trust members instead asked that the boys work in the garden as restitution. They like it so much that they got plots of their own to garden, and one of them became an officer of the Land Trust.

It is not easy for community groups to buy land in the Bay Area. Property here is very expensive. I've seen a quarter acre in San Francisco for sale for $1.1 million, and a hotel (acquired by the Chinese Community Housing Corporation to serve tenants like those who had lived at the International Hotel) valued at $500 a square foot. It is also very difficult for non-profit groups to get financing for properties. Traditional financing sources are notoriously skittish about cooperative ventures, and public funding is scarce. But land trusts do successfully acquire land even today.

In order for community groups to acquire and control land it is necessary for them to master the same technical skills that the real estate interests use. They must know how to incorporate and get tax exempt status (which enables them to offer tax advantages to donors) how to research property ownership and negotiate effectively.

It is exciting to see people working together to reclaim and improve their environments. Individuals banding together to own land is an American tradition. Eleven white men and seven black men met in Tyronza, Arkansas in July, 1934 to form this nation's first truly integrated union. They held a Ceremony of the Land in 1937 and they said, "Speed now the day when the plains and the hills and all the wealth thereof shall be the people's own . . . " Working together, we can make their dream a reality.

The I Hotel

When I finally ran through the cordon
with the others on the sidewalk
and reached the picket, there was
a sledgehammer sounding
and into the mural painted on
the Jackson St. Gallery wall
the cops were smashing for a doorway.
And when the crowd jeered with rage
at the desecration of that mass mural
I'd passed 100 times and 100 times
it had called me back to North Beach,
and the cops then called for reinforcements,
goons with nightsticks who lined up
along the wall and were still standing
ten minutes later along
that muralled wall painted by a group
of young Asian-American artists
in honor of their modern heritage
of ancient memory in revolution—

DOWN WITH FASCISM

I am still wailing, as if we too had been beaten
in, like Ray, earlier, staggering along
the sidewalk, or that painted door
splintered and torn out of the mural's landscape
like a wound I dip my brushes in
and write.

LONG LIVE THE I HOTEL

—Jack Hirschman

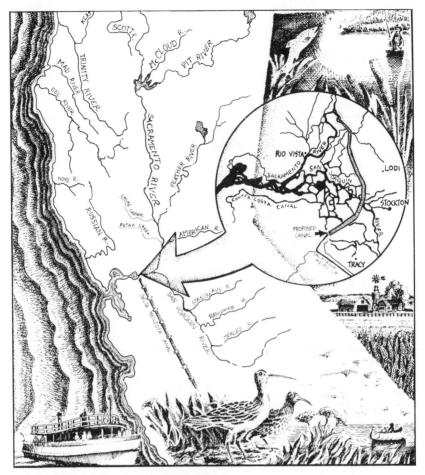

WATER PIRACY:
The Rise of Hydraulic Society and the Destruction of Northern California

Michael Helm

For the past 30 years the northern half of California has increasingly been converted from a land of awesome and diverse beauty—with giant forests, exquisitely unique rivers, deltas, and bays, abundant fish and wildlife, and thousands of family farms—into a support system for a highly centralized, hydraulic society managed in the interest of an economic elite headquartered in the southern part of the State. An economic and political elite dominated by the Chandler family of the Los Angeles Times, Big Oil, Agri-business, and other large corporate land developers and contractors.

The fountainhead of this transformation has been and continues to be massive innocuous sounding water development projects like the Owens River Aqueduct, the Central Valley Project, the Mono Lake Aqueduct, and the State Water Project. The southward diversion of water made possible by these projects has been the single most important ingredient in determining demographic, economic, political, and environmental trends throughout the state.

Currently before California voters is a referendum on S.B. 200, the $12-20 billion expansion of the State Water Project authorized by the state legislature in 1980. S.B. 200 includes the ominous and vampire-like, Peripheral Canal. A canal that many believe will spell the death of San Francisco Bay, the Sacramento-San Joaquin River Delta, and ultimately, California's North Coast rivers. If S.B. 200, or its equivalent in future years, is approved by California voters, we will see the further expansion of what already is the most massive and expensive water development project in the history of the planet.

As energy costs escalate over the next twenty years (caused in significant part by the energy consumption of these water projects, themselves) we are likely to see in California (and throughout the Southwest) the increasing emergence of a corporate, plantation agriculture sustained by publicly subsidized cheap water and the massive re-introduction of exploited "bracero" labor from oil rich but water starved Mexico. We are likely to see—at first imperceptibly, then more obviously—an increasingly repressive society in which millions of immigrant laborers work at subsistence wages with no political rights while the quality of urban life in both north and south deteriorates and northern Calfifornia's environment is sacrificed. Which way the water flows will be the single most important issue in determining the quality of all our lives in the 1980s and beyond.

Historical Background
Owens Valley and the Los Angeles Aqueduct

Back in 1904 Harry Chandler—of the *Los Angeles Times*—along with business cronies like Moses Sherman, Otto Brandt, and Isaac Van Nuys realized that speculative growth and profits in semi-arid southern California, and specifically in the parched San Fernando Valley, were dependent upon pirating an external water supply. Coincidentally, William Mulholland—head of the Los Angeles Water Board—was thinking about new sources of water, too. Mulholland took a trip some 250 miles northeast of Los Angeles to the Owens Valley, the site of a thriving agricultural community with over a quarter of a million acres of irrigated ranches and farms. What Mulholland saw, he liked; lots of fresh mountain water cascading down into the Owens River.

Once the source of additional water was identified, however, a problem still remained. How to get it to Los Angeles, and how to get the people to finance it? Mulholland and Chandler solved their predicament by fabricating a drought. The editorial and news sections of the *Times* became obsessed with the notion of an imminent water shortage. *Times*'s readers were propagandized into believing that from 1895 to 1904 the average rainfall in Southern California had only been half of normal. Los Angeles's local ground water supplies, readers were warned, were about to be depleted. In truth, the underground basins were full and capable of providing for a growth of up to one million people with prudent management and use—a ten fold increase in Los Angeles' 1900 population of around one hundred thousand inhabitants. But Chandler and Mulholland were thinking bigger. They were laying the foundation for a southwestern empire that ultimately, with its aqueducts, would become the new Rome. Successfully terrified by the *Times*'s drought

The Hydraulic Elite

(Those who get subsidized water from the State Water Project) —1980—

	Corporation	Subsidized Acreage		Corporation	Subsidized Acreage
Kern County	Tenneco	53,000	**Southern California**	Irvine Ranch	77,000
	Chevron	50,000		Newhall Land Co.	80,000
	Getty	42,404		Bixby Ranch	30,000
	Southern Pacific	38,000		Mission Viejo	20,000
	Shell	31,464			
	Tejon Ranch	22,000			
	Tejon Agricultural Partners	13,000		**Major Contractors**	
	Berrenda-Mesa	13,000		Fluor Corp.	
	Superior Oil	7,716		Southern California Edison	
	Union Oil	3,979		Pacific Lighting Co.	
	Blackwell Land Co.	22,430		Pacific Mutual Insurance	

Note: Besides the dollar value of subsidized water, the land value of many of these acres has increased from $200 to $2,000 per acre as a result of the State Water Project.

stories, Los Angeles voters approved a $25 million bond to build a gravity-flow aqueduct 232 miles into the Owens Valley along the eastern escarpment of the Sierra Nevada. With the aqueduct approved, Chandler and his friends formed a syndicate and quickly bought up 60,000 acres in the San Fernanco Valley. Without water the land was relatively worthless. But with water—especially publicly subsidized agricultural "surplus" water—the syndicate stood to reap over a hundred million dollars from their initial investment of two and a half million.

By 1913 the Los Angeles Aqueduct was completed and began diverting five times the water that Los Angeles residents could possibly use. At first, however, the San Fernando syndicate couldn't profit from it because the federal government had gotten wind of their scheme and had prohibited any of the Owens Valley water from being sold as surplus outside the city limits of Los Angeles. Ever resourceful, the syndicate overcame this temporary snag in their plans by embarking on a long series of artificial annexations to the city—beginning with the San Fernando Valley. Thus legally qualified, the syndicate began receiving hundreds of thousands of acre feet of drought-contrived surplus agricultural water that urban Los Angeles residents financed through repayment of the bond.

By 1926 the Owens Valley—its water diverted—had largely become a man-made alkali desert. When outraged Owens Valley farmers and ranchers reacted to this vampiric process by blowing up Mulholland's aqueduct, his imperial reply was to suggest that, "there aren't enough trees in the Owens Valley to hang its people on." Instead, Mulholland and his Los Angeles Department of Water and Power bought up a checkerboard series of plots of land in the Owens Valley and through a series of pumping operations lowered the groundwater level below the root structure of the plants that sustained the valley.

As the Owens Valley died, the San Fernando Valley blossomed along with Harry Chandler's bank account. The value of an acre there increased from twenty to two thousand dollars. By the 1930s, according to Robert Gottlieb and Irene Wolt's book, *Thinking Big*, "Harry Chandler owned more than 2 million acres in ranches, agricultural property, cattle and cotton operations, and suburban land. He had become California's premier land baron, an American version of the state's earlier establishment figure, the Spanish don." Among Chandler's holdings were included the San Fernando acreages, the 300,000 acre Tejon Ranch, 1,000 acres in the Imperial Valley, Rancho Santa Anita, the 350,000 acre Vermejo Ranch along the Colorado-New Mexico Border, and a large interest in the 860,000 acre C-M Ranch just south of the Mexican border. In addition, Chandler was intimately involved with Warner Brothers Studios, Union Oil, TWA, Douglas Aircraft, Yosemite National Park Company, Firestone Co., the Los Angeles Steamship Company (LASSSCO), and a host of other corporate enterprises.

In 1940 the Los Angeles Aqueduct, through a $38 million dollar bond, was extended from the Owens Valley to Mono Lake, some 338 miles from Los Angeles to a point north of San Francisco. By 1979, due to almost total stream diversions, Mono's water level had dropped over 35 feet, its water was increasingly saline, the California Gull's rookery was in danger of extinction, and alkali dust storms swirled their polluted contents for a radius of sixty miles.

The Colorado Aqueduct
and the Metropolitan Water District

With the Owens Valley water coming to Los Angeles, more people came, too, Los Angeles, with its annexation fever and growing suburban sprawl, was sold like a magic elixir. First the San Fernando Valley was farmed, then gradually subdivided. What had worked so profitably once was tried again. Los Angeles's emergent ruling class—led by the *Times*—cast its covetous eyes toward the Colorado River. If the Colorado were to be tamed and diverted, they reasoned, spectacular subsidized profits would accrue to their arid holdings in the Imperial and Coachella Valleys as well as in San Diego and Orange Counties.

In 1928, crying impending drought once again, the Chandler-led retinue helped form the *Metropolitan Water District* (MWD) which was dominated by the Los Angeles Department of Water and Power. The MWD, along with water interests in the Imperial and Coachella desert valleys, successfully lobbied the Roosevelt Administration into building giant Hoover Dam. In 1931 the MWD-*Times* axis successfully propagandized Los Angeles voters into approving the $220 million Colorado Aqueduct.

By 1941 the 242 mile long Colorado Aqueduct was completed and, along with the All American and Coachella Canals, began delivering what would ultimately reach 5.3 million acre-feet of Colorado River water to southern California. Of this amount, eighty percent would go to Agri-business —primarily in the Imperial Valley—to irrigate 675,000 acres of crops picked by Mexican laborers on land owned by the Chandlers and their corporate friends. The remaining twenty percent (1.2 maf) would go to MWD. But almost all of this MWD water went, not to Los Angeles, but was wholesaled at surplus rates to agricultural and suburban land developers in Orange and San Diego Counties. The prestigious *California Water Atlas* has estimated that from 1942-1972 alone, Los Angeles residents paid $335 million, mostly in hidden property taxes assessed by MWD, but actually received only 8 percent of the water that money theoretically entitled them to.

By 1961 the Colorado River was so developed that it had ceased to run into the Gulf of California. The last 1.5 million acre-feet were used—in degraded saline form—by Mexico before they reached the sea. With Federal and Los Angeles subsidized Colorado River water, southern California's population increased from 3.3 million in 1940 to over 10 million in 1970.

The State Water Project

What worked so well with the Owens and Colorado River aqueducts was rolled out once again in 1960. Screaming that southern Californians were going to run out of water and that Kern County agri-business was threatened by groundwater overdraft problems, the *Times* succeeded in convincing enough Los Angeles voters so that the $1.9 billion Phase 1 of the State Water Project was approved—statewide—by a scant 130,000 out of a total 5 million votes cast. What the *Times* didn't tell voters was that their 300,000 acre Tejon Ranch along with a handful of oil companies in Kern County would be the principal beneficiaries of diverting some 2.3 maf of water from the Feather River in Northern California via the building of the Oroville Dam, the Clifton Court Forebay pumping plant in the Delta, and the connecting 444 mile long California Aqueduct. The *Times* didn't tell voters that Agri-business interests were pushing for the State Water Project because it would enable them to receive cheap subsidized water without the 160 acre restriction theoretically imposed by the Federally operated, Central Valley Project.

Tejon Ranch and Southern Pacific were the two largest financial contributors to groups pushing for construction of the State Water Project (SWP) in 1960. In addition, the Arvin Rock Company, a Tejon subsidiary, provided most of the materials for construction of the aqueduct. Though the *Times* had predicted that southern California's population would bulge to 22 million by 1970—thus justifying the need for the State Water Project—in fact the charm of uncontrolled growth was beginning to wear thin. By 1980 the total southland population had barely creeped over 12 million and Los Angeles was actually experiencing a decline.

In terms of water received, twelve years after Phase 1 of the State Water Project was completed in 1968, Los Angeles residents were only receiving 18,000 acre-feet of water out of a paid MWD entitlement of 540,000 acre-feet. It was the old surplus agricultural scam once again. Almost all of the SWP water that Los Angeles people were paying for was being sold to the Kern County Water Agency who in turn sold it at fourteen cents on the dollar to a handful of oil-agribusiness corporations including Chevron, Getty, Tenneco, Shell and Union Oil, as well as Southern Pacific and the Chandler owned J.G. Boswell and Tejon Ranch properties.

In 1980 over one half (1.3 out of 2.5 maf) of the entire yield of the State Water Project went to Kern County oil-agribusiness. Half of that went at surplus rates of $3.50 an acre-foot (instead of the normal $23) and was used primarily to grow cotton on 150,000 acres of reclaimed desert land. Very little of it went to correct groundwater over-draft in Kern County as previously promised.

Phase II of the State Water Project

Greed apparently knows no bounds. Having already drained the Owens and Colorado Rivers, Mono Lake, and 2.5 million acre-feet out of the San Francisco Bay-Delta Estuary, southern California's hydraulic elite thirsts for even more. Specifically, it thirsts for the $14-20 billion S.B. 200 which includes the massive 43 mile long, 400 foot wide, and 30 foot deep Peripheral Canal. A canal which will allow the diversion south of an additional 1-2 maf of Sacramento River water. The Peripheral Canal, once in place, will also make it possible to later drain the North Coast rivers, the Eel and the Klamath.

Though the MWD is pushing for voter approval of S.B 200, the truth is that according to its own internal documents urban southern California does not need any additional water. Many analysts, including the Rand Corporation, are arguing that—even if the Peripheral Canal is built—the water and the energy required to deliver it will be so expensive ($1.5 billion annually) that southern Californians won't want to pay for it. Many believe that S.B. 200 is actually designed to *guarantee* further subsidized, surplus water for Kern County oil-agribusiness expansion, to guarantee that Orange and San Diego County land developers will continue to be subsidized by Los Angeles taxpayers.

Certainly those lobbying for S.B. 200 give credence to this view. The MWD—with its $200 million annual budget and 1,300 employees—spent over a million dollars in 1980 promoting the Peripheral Canal. In 1981 MWD is the second largest legislative body in California. With its political muscle and huge budget it has become, in dissident MWD member Ellen Stearn Harris's words, "the *de facto* land and development planning agency in California."

Interestingly, none of MWD's 52 board of directors is elected; rather they are appointed—in some cases for life—by the member water districts that MWD wholesales its water to. Over half of MWD's board members, according to the *Sacramento Bee*, have substantial investments in land development, real estate, banking, insurance, construction and engineering which would financially benefit from the importation of more northern California water. Many of the corporations that represent these investments banded together, with MWD encouragement, to form a political action group called *Californians For Water*. This group, which successfully lobbied S.B. 200 through the state legislature in 1980, was funded by large ($10,000) contributions from such corporate donors as Getty and Union Oil, the Newhall Land and Cattle Company, Bixby Ranch, Irvine Ranch, Southern California Edison, Pacific Light, Security Pacific Bank, Mission Viejo Land Co., and the Fluor Corporation. It is a reasonable surmise that these corporations expect substantial benefits from S.B. 200.

Californians for Water and MWD—supported editorially by the *Times*—are basing their arguments for more of northern California's water on the assumption that southern California will lose its right to 700,000 acre-feet of water from the Colorado River, beginning in 1985, due to a 1964 Supreme Court decision apportioning that amount to Arizona. That may or may not be true, depending upon whether or not Arizona takes its full share, and what the total runoff into the Colorado River watershed is in future years.

What MWD isn't saying, however, is that—at far less expensive and environmental damage than is likely from S.B. 200—the southland could more than make up for any Colorado River loss by paying to line the Coachella and All American Canals (thus saving 500,000 acre feet) and by initiating a quite feasible 15 percent conservation and reclamation program. The reason these alternatives are being ignored is they would eliminate the surplus water subsidy that the special interests in Orange and San Diego Counties have been getting for the past forty years.

Environmental Costs of the Peripheral Canal

Bad as the economic and energy costs of S.B. 200 and the Peripheral Canal are for southern California, the effects on northern California are likely to be catastrophic. The MWD claims that it only wants the Peripheral Canal in order to efficiently tap the "excess" spring-time water that flows from the Sacramento River through the Delta into San Francisco Bay and then out to sea.

But the truth is that there is no excess water available. In 1980 *two-thirds* of the historic fresh water flow into the San Francisco Bay-Delta Estuary is already diverted from the Sacramento-San Joaquin River watershed system. Primarily by the Federally operated Central Valley Project (6.5 maf), and by Phase 1 of the State Water Project (2.5 maf). Taking any more water out of the Delta would be like prescribing a blood-letting for a patient suffering from anemia. The San Joaquin River is already developed to the point that it actually has a *negative* flow during part of the year. At present 88 percent of the fresh water flow into the Delta comes from the Sacramento River.

If the Peripheral Canal is built, with its capacity to divert 70 percent of the Sacramento River's flow south, the 738,000 acre Delta farmland—worth $350 million a year—will be endangered. The reason for this is that with increased diversions of fresh water from the Sacramento River south, the salt water tides that diurnally come in from the Golden Gate in San Francisco Bay will likely intrude far enough into the Delta to contaminate the fresh water needed for irrigation by the farms that reside there. In the drought year of 1977, without the Peripheral Canal's increased diversions, salt water intruded nearly 80 miles into the Delta as far as Stockton.

In addition, the hydrology of San Francisco Bay is poorly understood. Many believe that, with any further reduction of the fresh water flows from the Sacramento River, south San Francisco Bay will no longer be adequately flushed—that it will permanently turn into a noxious smelling, chemical sump. Given the likely increase in Bay Area urban growth, a strong case can be made for requiring *more*, not less, fresh water flow into San Francisco Bay between now and the year 2000 just to maintain existing water quality.

While the proponents of the Peripheral Canal claim that it will help the Salmon and Striped Bass fishery in the Delta, U.S. Fish and Wildlife disagrees. The huge pumping plants in the Delta near Tracy which create reverse flows—thus confusing spawning fish—will not be totally eliminated by the direct tap into the Sacramento River of the Peripheral Canal. The reason for this is that the Central Valley Project's pumps will still be annually pumping over 3 million acre-feet of water out of the Delta south. Then too, the $120 million fish screens that are to be placed at the head of the Peripheral Canal are unproven and will probably kill millions of the delicate fry.

But the largest truth about the fishery is that already 65 percent of the recent historic salmon and striped bass population has been decimated since 1951. Studies by the Fish and Game Department have indicated that fresh water diversions have been the single most important factor in reducing the Delta-Bay fishery. It is insane to believe that the Peripheral Canal will improve the fishery when the main reason for building it is to increase the total volume of diversions.

Even further is the historically justified fear of North Coast people that what really lies at the base of the Peripheral Canal's ultimate design is the future theft of the waters that flow into the Eel and Klamath Rivers. Already the north fork of the Trinity River has been pitifully drained and dumped into the Sacramento River. With the Peripheral Canal in place, and given the unending thirst and political power of the Southland, what is to prevent the Eel and Klamath from being grabbed in the future? As the Delta and San Francisco Bay are further degraded by the increased diversions that the Peripheral Canal will effect, it is likely that these areas will be forced to sell out the North Coast rivers in order to survive.

The solution to maintaining and enhancing the Delta and San Francisco Bay is not further diversion of northern California water, but rather to institute sound water conservation policies in the San Joaquin and Sacramento Valleys. Currently 85 percent of the diverted water from the Central Valley Project and the State Water Project goes to agriculture which squanders it with inefficient open ditch irrigation. If the state were to take part of the $20 billion price tag for S.B. 200 and the Peripheral Canal and subsidize conversion to sprinkler and drip irrigation, perhaps by as much as 33 percent—or 3 million acre-feet per year of northern California's developed water could be allowed to flow—as nature intended it—into the Delta and San Francisco Bay once again. A portion of this saving could also be used to replenish the badly overdrafted groundwater supplies in the Central Valley.

Another benefit would accrue from a properly restored Delta and Bay. Preliminary studies by U.S. Fish and Wildlife have indicated that over a billion dollars a year in fishery, agriculture, and recreation values could be generated there with proper reclamation. Does it make even economic sense to abandon this potential so that 300,000 subsidized acres of desert land—worth $188 million—can be farmed by oil companies in Kern County via the Peripheral Canal?

Energy

But the expansion of the State Water Project via S.B. 200 is a mistake in another sense, too. The State Water Project was originally built on the assumption of perpetual cheap energy that would allow for the economic pumping of diverted water south against the natural flow of gravity. Those days are now gone. The price of a barrel of oil has increased from $2 a barrel to over $30 a barrel, and the end is not in sight. All the best hydro-electric sites have already been developed in the state. If Phase II of the State Water Project is built, its demand for energy will increase from 6 billion kilowatts to 11 billion per year. The SWP, which is already the single largest user of energy in the state, will increase its share of total energy consumption from 8 percent to perhaps as high as 20 percent. More environmentally damaging coal or nuclear plants—at an estimated $5 billion cost—will have to be built to provide the extra capacity. Though large agri-business and industrial users will primarily benefit from this increased capacity, urban homeowners and renters will pay for 70 percent of the cost.

Future Scenario

If the Peripheral Canal and its attendant reservoirs and power plants are built, California will have taken one more giant step towards being a totally hydraulic state. One in which technocrats in Sacramento, by pushing a few computer buttons, can move water at will throughout the state according to the bidding of the powerful southern California oil-agribusiness and land development elite. But it won't stop there. These interests already have contingency plans to tap the Snake River. The Federal limitation of 160 acres for subsidized water is about to be gutted by them. They've seriously proposed building a North American Water and Power Alliance (NAWAPA) that would divert water from as far away as the Yukon and distribute it to points throughout the southwestern United States. Ronald Reagan, the spokesman for Southern California's elite, has indicated that Federal water development in the West will be given renewed priority.

As energy costs rise over the next twenty years there is going to be an increasing strain on California's $14 billion agriculture to remain competitive. The energy costs of processing and transporting its products over thousands of miles will—unless further subsidized—become prohibitive. The energy costs of fertilizers, pesticides, and of operating farm machinery will also significantly increase.

Anticipating these trends, the hydraulic elite is clamoring for a renewed source of cheap, foreign labor. President Reagan has already approved the first stage of a renewed "bracero" program for California, Arizona, and Texas. This program is cosmetic and will not stem the tide of the estimated 1 million "illegal" workers who cross the Mexican border each year. Millions more will be coming. They will migrate north to till the corporate fields at subsistence wages and to serve as domestics and cheap, service-industry laborers in the cities. California will increasingly become an exploitive, neo-feudal society. A society that will sooner or later require virtual militarization in order to deal with the inequities and anger that it will engender.

Ultimately, of course, southern California—like all hydraulic, reclaimed desert societies of the past—will stumble and fall of its own weight. Sooner or later the land will no longer be able to sustain the introduction of massive amounts of water and people to where nature didn't intend them. The land will gradually—as is happening even now—accumulate minerals and salts from the imported water. The crops will begin to wither and die. The desert will reclaim its own. Northern California, its resources extracted, will lay waste. It needn't be this way. Decentralist alternatives that are ecologically and politically sound exist. But, it is getting very late in the imperial game to introduce them.

In Case of Illumination

in the event of fire
break glass
& operate switch.

over.

dying is hazardous
to your health.

stop.

in case of fire
at night
do not use normal exists.

over.

if only there were a drug.

stop.

the half-life of plutonium
is 5 million years
as far as we can determine.

over.

prenatal memories
found to be primary
source of neurotic anxieties
cancer
visions.

stop.

the half-life of uranium 235
is 10 million years
by our calculations.

stop.

don't panic.

continued

in case of illumination
fold hands together.

radiate.

over.

the half-life of life
is so far
undeterminable.

stop.

thought atrophies
at the rate
a cell explodes.

over.

take 2 thorazine every hour
on the hour.

stop.

fire escapes.

the air divides again.

over.

words are guilty
of fusion.

the mouth catches
radiant butterflies that disappear.

over.

there is nothing
one does that is not lost.

stop.

the half-life of desire
is exactly.

—*Cliff Eisner*

ON BEING OUR OWN ANTHROPOLOGISTS

Peter Berg, talking

I want to talk tonight about *place*-locatedness. Recently a woman from Germany visited me who teaches a course on American popular culture at the University of Frankfurt. Every summer she goes around the United States dutifully collecting *data* about American popular culture. And it was strange talking with her, anyone in this room now would have felt the same way, because I was suddenly a representative of American popular culture. I didn't know whether to offer her a beer, joint, coffee, or organic strawberry juice, right? So I began thinking about how synthesized American culture is, wondering what things were deep-seated, long-term parts of American Civilization.

So tonight, I would like for you to imagine a Balinese, someone who has been studied to death by Western anthropologists, say a Balinese dancer in her early 20's coming here to study American culture. She's a third generation of the Balinese who have been minutely documented. Gregory Bateson photographed her grandmother breast-feeding her baby. Her father was filmed doing a monkey dance by a BBC documentary team. When she was a kid her toilet habits were studied by Margaret Mead. So her family has impressive credentials as subjects of cultural anthropologists. But, none of these Balinese subjects know where these people who were studying them came from. So let's say that in an incredible bureaucratic mix-up at the U.N. our Balinese dancer is given a grant to come to the U.S. to study American culture for the Balinese.

She arrives at the L.A. airport, right in the middle of American global monoculture. She's given a rented car with a chauffeur, gets a suite at the Hilton Hotel, and then goes out to find Americans, to find out what their deep seated culture is. And the people she runs into on the street are gas station attendants, shoppers, somebody distributing the Watchtower, and so forth. She wades through all our technological garbage to find people to talk to. Every time she goes across the street there are lights directing her. There are more machines than she's ever seen. Everybody relates to machines. Everybody seems to be dominated by these machines.

Our Balinese dancer wants to find out what the people have in common, the roots of their culture. So like a good cultural anthropologist she asks them what they call the days of the week. They say Monday, Tuesday, . . . She finds out that this is Norse Mythology! *Moon* day, *Tuwaz's* day—the one armed hero's day, *Woden* day, *Thor's* day, a day for *Frei*—the goddess of fertility, and *Sun*day—the sun's day. So, all our days are either named after Norse gods or planets. This sounds a little like Bali to our visitor.

And then what about the months? She finds out that they start with *Janus*, the two faced god of the Romans. And the rest are mainly drawn from Roman roots. Some of them are named after Roman numbers; *Sept, Oct, Novem*. She detects that there was a Norse world of gods and planets that was overthrown or displaced by Romans. And the years? The years are an accumulation after the birth of an Aramaic-speaking Judean prophet!

So, can you begin to see that there is something about this society, about its *dislocatedness*, the Super Society, the media society of the United States, Canada, North America, the West generally, that's built to slide? It's been transformed so many times that we've got a barely remembered pantheon of Norse gods that the Romans knocked out and now we're accumulating numbers after the birth of Christ toward what? Toward Judgment Day! Because it is officially ordained that we will end as a species; the world will end and we are keeping a count down of the years before that happens. Our civilization is an accumulation of dislocated and displaced cultures. It's really not the product of a progression—"the ascent of man." It's not really a triumphant march upward. Some people on the planet still have cultures that are place-located like the Balinese. The things they see around them resonate with their culture. But American, Western culture is set up in a spirit of transformation. *Transformation* is its dominant theme.

Beware of futurists. That's what I'm coming to. But first, I want to work up to that from some other aspects of transformation. For example, our science is transformative science. There are, of course, other ways of thinking which are integrative. But our science takes things apart to change them. We are compelled to feel that we should *change* everything. Change for its own sake.

There is another sense of time, too. The other sense of time is *now*ever time. There are people who think of time being an ever-present *re-creation* of everything that's gone before, and everything that will come out of that. *Now*ever time, cyclic time. Our time is linear, march of time, and has an apocalyptic edge built into it, like Judgment Day.

So our civilization, our culture, whatever you want to call it, is an accumulation of dislocated cultures with a transformative, apocalyptic point of view built into it. Why else would someone believe that if they put themselves into a cylindrical metal tube and were shot *out* of the planetary biosphere that that would be a superior challenge to anything that would be met *in* the biosphere? Why would someone believe *that* unless they were already convinced of the value of change for its own sake? That because there was noplace that you started from, it really didn't matter where you went. In fact, the bigger the odds, the better. It's like the search for the Golden Fleece or looking for the chalice that Christ drank from. An incredible mission. Mission Impossible. Getting out there. Ironically, there is no *out* once you are in outerspace. You're trapped in a metal tube.

What I'm going to submit is that people don't really need or want that. As a matter of fact we'd have to consume a tremendous amount of people's labor, a tremendous amount of energy, and have to hype people into somehow believing they needed or even really wanted to be shot out into space. People have been so coerced by constant transformation, technological transformation, that they feel going into space is a natural extension of that. And, in fact it is a natural extension of transformation. But is it worth it?

The people who have come to North America generally have not cared so much about what was here as for what they could do with it, how they could use a lot of technological toys to get things out of it. Western culture has largely come under the domination now of American culture; and what it tends to represent is something I've called *global monoculture*: Sitting in a gas line in a Toyota with a decal on the back that says, *Save the Whales*, for four hours reading 'Jonathan Livingston Seagull.' What an incredibly ironic place for people to end up at!?

I think the futurists are about to take the future, remove our sense of *now-ever*, and I want to speak against that. I want to speak in terms of a *culture of resistance*. The corporate involvement that goes into space pushes the idea of compressed urban populations. It also has a military basis and works hand in hand with the military. There is a kind of theft going on here and I believe that part of the theft is this: we have a possibility to begin designing a vision of what human beings are that takes in all of the peoples of the planet in a way that is in harmony with the needs of the planetary biosphere. We have a concrete basis for fulfilling those kinds of wishful, idealistic urges that take the form of wanting, for example, to abolish wars. We have a place for conceiving our way now. We've all seen that image of the planet and realize that we don't have to be hooked into killing a lot of other people just because they don't happen to belong to the national group that we belong to. There is the possibility of seeing human culture, the human species, as relating to the rest of the biosphere. This seems to me to be a proposition on the scale of wanting liberty in the 18th century, when people had aspirations to get, for the first time, the vote. We have the possibility now of thinking in terms of what the human species might actually do to maintain and restore the biosphere, to begin actively living in continuity with it. I don't want to have that robbed by futurists and I don't want to be bamboozled into thinking that our sweat and energy necessarily has to go up in space.

So what would a culture of resistance entail? In Northern California, which I think is a distinct country of the planetary biosphere, a cultural resistance might be formed by sharing views of the place itself, about the web of life that sustains it and us. To be more specific, let's divide Northern California into four distinct locales: urban, suburban, rural and wilderness. The urban areas might be thought of as "Green City". San Francisco could have half the streets it now has. Either because half of them were closed and dug up to provide topsoil, or because the number of lanes on all of them were divided in half and one half of the streets were dug up for topsoil so things could be planted in them. A real civic effort would be made in support of the kinds of things we've already started like neighborhood co-ops; that neighborhood co-ops could in fact be given the kinds of advantages that businesses have always been given. I think it could be measured whether or not a community store is of value to the neighborhood, and, if it is, then it seems to me that some tax money from San

Francisco could be diverted into supporting that endeavor, rather than having people exhaust themselves to keep co-ops floating. The kinds of culture that are presented in the Bay Area might be more overtly cultures of the North Pacific Rim. The kinds of theater and art being done outside of San Francisco could be brought into the city so people could see real examples of people taking over their watersheds in the Salmon River Valley, in the Mattole River Valley, things that are *really* going on.

In the suburbs, a "Green Region" plan might insist that if you're going to live there you have to involve yourself in some kind of part-time agriculture. This would be a way of balancing the kind of pressure that suburbanites put on the landscape by covering up the topsoil and using tremendous amounts of energy to commute back and forth to these places. It wouldn't have to be too much more than learning to provide some of the produce for your own family rather than aping the rich with croquet lawns.

The deeper rural areas of Northern California produce farm products on a large scale and the coasts have fisheries that shouldn't have to compete with rainbow trout from Peru. Rather than producing for the global monoculture our economy would be more regionally oriented and symbiotic. We would clean up our rivers, restore watersheds, and perhaps become Salmon people in the way that we see the place.

For the wilderness thing, we need to overcome the Sierra Club psychosis. The Sierra Club psychosis is that someone went into the woods once and had a tremendous experience and never wanted to have that experience taken away. It's not unusual now for some people in the Sierra Club to talk about the desirability of nuclear reactors. Nuclear reactors would be good because they would centralize human populations and keep people out of the wilderness areas. That's close to psychosis. The Sierra Club supported the Peripheral Canal which would have diverted half the water from around the Delta, turned San Francisco Bay into a sump, and transported that water to replace the depleted ground water in the Central Valley and to Los Angeles. One of the reasons for supporting the Peripheral Canal that I heard from a Sierra Club board member was that it was good to keep the people in Los Angeles because if we kept them there they'd be kept out of the wilderness areas of Northern California. It's crazy to destroy our watersheds to keep people in Los Angeles.

So what we really need is some kind of unified vision of what a Green Region, a re-inhabited region, would be and what our civilization, our culture, might be in terms of this place of the planetary biosphere. Then when that woman from Bali comes to visit us she can be told that she is in *Shasta*, that the people that live here are involved in the migration of natural species that occur here, that they have a feeling of fairness and sharing, that they are undertaking programs to secure the long-term inhabitation of this place, and that they have a culture of resistance against the Global Monoculture.

Raise the Stakes!

What is it worth to polish off condors, Amazonia, the Bretons, the Pacific . . . our own great grandchildren?

What are we asking for it? When airfield-long Japanese lumber barges shovel Amazon forests into dust? When machine-gunning helicopters find the last unknown native tribes?

(Even though those trees put out plenty already. Oxygen. A lung of the planet-body. And the natives are the only people who know how to live there.)

Nothing is worth it. Natives, Narita Airport diehards, oilspill-drenched Bretons, the pregnant women of Three Mile Island—are us. We do know each other: nobody's childhood place looks the same anymore.

Raise the stakes. We are the same species. We don't come from . . . where? We come from everywhere.

Raise the stakes to, WE'RE SPECIES-KIND. WE COME FROM EVERYWHERE.

We come from everywhere and we all live some place. National interest? Multi-national interest? They move around too much. Never close to home. They don't really have a place.

The planet is alive and places are alive. Places are multi-species ground and people are a part. Raise the stakes to, PLACES ARE ALIVE. That's planet interest.

Raise the stakes. Higher pay isn't enough. Shorter jail sentences aren't enough. The government might fix your teeth? It isn't enough. We're killing too much to be able to live.

Raise the stakes. We have the right to defend ourselves as a species. Planet-local and planet-wide. We have the right to be culturally diverse; as different as the places where we live. We have the right to govern our planet-places. We have the right to agree on our role as a species sharing life on this planet with other species.

How do we secure those rights? How do we make that agreement?

Let's raise the stakes and find out.

—*Peter Berg*

RECONSTITUTING CALIFORNIA

Jack Forbes

I am suggesting that we adopt an initiative which would provide for the possible division of California, subject to these new entities being admitted to the national union by Congress.

Very briefly, I am proposing that the voters of this state should possess the right to divide California as an inherent right of democratic self-government, provided that it be done in some kind of orderly, logical manner.

To achieve this we could
divide California into a series of regions, as follows:

PALOMAR
(San Diego and Imperial counties);

RAMONA
(Los Angeles, Orange, San Bernardino);

INYO
(Inyo, Mono, Alpine);

YOSEMITE
(Santa Barbara, San Luis Obispo, Monterey, San Benito, Kings, Kern, Tulare, Fresno, Merced, Mariposa, Madera, Tuolumne, Calaveras, and Stanislaus); and

SHASTA
(all of the remainder except that Riverside may join Palomar or Ramona, Ventura may join Ramona or Yosemite, and San Joaquin may join Yosemite or Shasta).

After the initiative was passed into law those regions along the edges of present California (Shasta, Palomar, and Inyo) would be entitled to form separate states (or to join Nevada, in the case of Inyo) after appropriate elections.

The voters of Ramona and Yosemite would have to wait and set up new states because their location precludes pulling out until after action by either Shasta or Palomar.

The proposal also provides the opportunity for Palomar and Ramona voters to form a single state if they wish and the same is true for Shasta and Yosemite. The objective of this plan, in short, is not to arbitrarily chart out the future of California in some rigid way but rather to retain a degree of flexibility.

I have not analyzed the proposed new states to see if Democrats or Republicans, Whites or Blacks, or Chicanos and Indians will gain any advantage. The reason for this "oversight" is that I consider it *inherently* advantageous for *people* of all races to possess legislative districts of near-neighborhood size. The number of ethnic minorities holding office would have to sharply increase under this plan without, however, making any special efforts in that direction.

Significantly, also, the following other advantages would accrue:

(1) campaign costs would be greatly lowered, thus allowing more people to run for office without becoming indebted to special interests;

(2) the cost of government, in general, could be made more consistent with the needs of particular communities;

(3) new counties would doubtless be created by the new states (especially Ramona and Palomar), thus making local government more rational and responsive (for example, breaking up huge Los Angeles and San Bernardino counties);

(4) new constitutions could also make possible the easier creation of

city-counties or other forms of organization which might end the duplication of overlapping city, county, and special district governments;

(5) similarly, fossilized school districts, such as that of Los Angeles, could be broken up and recombined with other districts that currently exist as islands or enclaves; and (perhaps most importantly);

(6) local people can have a meaningful role in determining the future of their homeland, of the region in which they live;

(7) the number of Federal Senate seats would increase from 2 to 10 senators, thus giving us a much stronger voice in Congress.

Many of us might have a vision of what we want Shasta or Yosemite to be like in 2000. We might still want the Eel and the Klamath to flow to the sea. We might want to see green hills between Oakland and Sacramento. We might want to fish in unpolluted streams, or ride bicycles along bike paths stretching from sea to mountains.

Whatever it is that we want for Shasta or Yosemite, at present, is not a dream but an illusion. Why? Because we do not have the majority vote over our own fate. Palomar and Ramona (southern California) possess veto-power over all our wishes, and that is a political fact of life.

But I don't propose to pit region against region. *All* parts of the state can benefit from division and it is in that spirit that this plan is offered.

Editor's Note—The specific amendment, proposed by Jack Forbes, can be obtained by writing to the Planet/Drum Foundation, Box 31231, San Francisco, CA 94131. Membership to this innovative foundation and its exciting publication *Raise the Stakes* is $10.00 a year.

The Bus Passed an Hour Ago

Tomorrow is against the wall
With its hands up.

We are in the middle of a robbery
And the word HOSTAGE occurs suddenly,
Like a revolver on the windowsill.

There is nothing to do
But freeze.

Then they take us to be tied
And pushed into the back room,

Where we will be lucky
To be found again.

—*Cliff Eisner*

Sympathy for the Devil

Maybe this is mad, he said,
but when is sympathy ever rational,
sympathy that matters?

Not too much, you understand, but enough
to fill out the man again (or should
I say, *mannequin?*)

and claim him as our own.

Listen, he said,
a culture is a living organism
like you and me,

and when we are ill as a people,
congested, corrupted, constipated, congealed, etc., the culture
may prescribe its own bitter pill
to purge itself to get well.

Well, Nixon was that bitter pill we had
to swallow to throw it all up: Viet-
Nam, Watergate, Koreagate,
Assassin-Nation,
Hate . . .

continued

Even his name is a curse
with that big warning X on the label, Nix
on you, Nix on me, Nixon US!
And he did, he warned us again and again
to see: "Make no mistake about this,"
he said; and, "I want to make
this perfectly clear,"
he said . . .

And for those with ears to hear
and eyes to see, Nixon unreeled a national
Hypocrisy so malignant, so deep,
we have, years later, still to face it
in ourselves, preferring instead

that "fictional" J.R. character on T.V.

The old sayings tell it all:
how the snake bites its own tail, how we
swallow illness to effect a cure,
"hair of the dog that bit you,"
and what about

that Biblical scapegoat with its bell
driven from the circle?
Rilke said, "Do not be bewildered by
surfaces; in the depths all becomes law,"
and our prescription read,

TAKE FOR TWO ELECTIONS, and we did.

And Nixon did his work well, agitating
the stomach, loosening the bowels, throwing
us up to ourselves, yes, *to ourselves!*

And make no mistake about this,
Nixon continues to work his culture's will.
So, a little sympathy for the devil
who hides and hides . . . only to reveal.

—Frank Polite

EDITOR'S POSTSCRIPT

Northern California is an unique and diverse region of the planet. A number of us, undaunted by New York/L.A.-oriented mass media cynicism, would like to keep it evolving in that way. Toward that end, **City/Country Miners** has presented the urban and rural weather of some of Northern California's most finely tuned minds. A weather that is larger, more subtle and complex, than the monocultural interests of the "Moral" Majority, a Southern Pacific clear-cut forest, or the 7 o'clock evening news.

Those of you familiar with the 15 issues of **City Miner Magazine** published between 1976 and 1980 will see that this book is an extension of that tradition. A tradition that won't tell you how to get rich quick or intimidate your neighbor, but might well open you to renewed possibilities both within yourselves and to the otherness around you. If you have *dug* **City/Country Miners**, I hope that you will tell your friends about it and contribute ideas and work to next year's annual anthology (deadline May 1st, 1982). There is a lot happening in urban and rural Northern California that we have just begun to scratch the surface of. The interdependence of town and country is something we will all have to increasingly take cognizance of. The Medfly spaying of the Bay Area's urban population, the Peripheral Canal debate, and the Diablo Canyon nuclear installation are but several recent examples that illustrate this fact.

This volume of **City/Country Miners** has featured urban and rural stories, poems, articles, letters, journals, oral histories, and graphic work organized around the four categories of *relationships*, *places*, *work*, and *politics*. The basic editorial bias has been to emphasize first person, direct, statement. I believe that, in grappling with the ever illusive nature of reality, many I's (eyes) are generally better than reliance on the royal, we, or third person "objective" voice. The naked I conveys only the authority that its own sense of authenticity can generate. Each reader, without editorial intrusion, has to decide for him or her self what rings true. My purpose in selecting the work which appears here has not been to establish any party line, but rather to present a diversity of indigenous voices. Voices which acknowledge both the torque of living in contemporary Northern California as well as some sense of vision about how we might evolve toward a greater understanding of ourselves and our environments. In addition, a good deal of historical information has been included. Past, present, and future are all interdependently connected to the wheel of time, and we ignore that connection at our peril.

Besides the article on Water Piracy, I am responsible for generating the oral histories featured in this edition of **City/Country Miners**. There is a wealth of personal experience and wisdom residing in the plain talk of people communicating directly with each other. I hope that some of you will contribute oral histories to the next edition of **City/Country Miners** drawn from unmined voices in your community. Especially, around the theme of interesting work.

I would like to thank all those who have contributed to making this edition of **City/Country Miners** possible and to encourage everyone to keep on digging. We have nothing to lose but our gloss.

—*Michael Helm September 14, 1981*

CONTRIBUTORS' NOTES

John Krich is the author of *Bump City: Winners and Losers in Oakland* (City Miner Books). He advises readers that any resemblance between the character in his story and himself in real life is purely coincidental. John has a hot new novel, *A Totally Free Man*, just out with Creative Arts. **Jim Dodge** lives on Roothog Ranch near Cazadero. Along with **Leonard Charles** and friends, Jim edits *Upriver/Downriver*, an excellent Northern California newsletter (Box 390, Cazadero, CA 95421—$6/yr). His prose story is reprinted from it, as well as Leonard's and **Linn House**'s. Linn and his cohorts up in Petrolia just received a modest grant from the Department of Fish and Game to pursue their salmon restoration project. **Nancy von Stoutenburg** is a graphic artist and designed *City/Country Miners*. **Jennifer Stone** lives in Berkeley and works as a film critic. Her journal entries are excerpted from a forthcoming novel, *Babes of the Bathwater* (New World Press). **Jeffrey Zable** lives in San Francisco. He advises you to immediately purchase copies of his books: *The Dead Die Young* (Androgyne Books), *Ashes Bear Witness To The Burning* (Ptolemy Press), and *Severed Branches* (Artaud's Elbow). **Daniel Roebuck** lives in Paris. In response to my query as to how he heard about us he writes, "I'm afraid the way I heard about *City Miner* was in a directory of little magazines . . . it said you were interested in city life which is in part how I see Situation No. 2." **John Lowry** lives in New York City. He writes, "Nancy said, 'See, they don't want New Yorkers.' I said, 'No, no, you have it wrong, they have a great sense of humor.'" **Thomas Farber** lives in Berkeley. His story is reprinted from *Hazards to the Human Heart* (Dutton). He is also author of *Who Wrote the Book of Love*. **Deborah Frankel** lives somewhere in Emeryville. **Arthur Okamura** lives in Berkeley as a refugee from Bolinas. He teaches at the College of Arts and Crafts in Oakland. **Ed Buryn** supplied most of the photos in this book. He is the author of *Vagabonding in America* (And/Or Press). **Brent Richardson** shares a warehouse with a bunch of artists in Berkeley. **Trina Robbins** lives in San Francisco and has had her comic work printed all over. **Joel Beck** doodles in Point Richmond. **Kristen Wetterhahn** lives in North Beach with poet **Jack Hirschman**. **Barbara Suszeanne** lives in Oakland and is a song writer. **Chris King** is a bilingual elementary school teacher in Geyserville. She is currently at work translating an anthology of contemporary Latin American poetry. **George Randall Griffin**, rumor has it, is a pseudonym. **Tom Hile** works as a gardener and pizza cook in Berkeley. He "never liked school and still stares out the window wondering what the question was." **Frank Polite** is author of *Letters Of Transit* (City Miner Books). He currently lives in Youngstown, Ohio, pursuing his negative capability and boche ball. **Lenore Kandel** lives in San Francisco and threatens to get organized one of these days. **Andy Brumer** is author of *Turtle* and lives in Oakland. **F.A. Nettlebeck**, at last word, lived in a van in Santa Cruz. **Alta** has her *Shameless Hussy* (Crossings Press) just out. **Bruce Hawkins** is a mainstay of the Berkeley Poets Cooperative. **William Garrett** is the author of *Jazz Piano On A Red Mountaintop* (Poetry for the People, SF). He runs Ampersand Typography and Artaud's Elbow (press) in Berkeley. **Ray Dasmann** is a bio-geographer, noted scholar, who is a Provost at U.C. Santa Cruz. **Susan Pepperwood** "lives with her five year old son on a knoll overlooking the junction of three creeks, five miles from the nearest paved road in the foothills of Mendocino. When she is not organizing the local community, she is planting even more flowers around her house." **Malcolm Margolin** lives in Berkeley and has a new anthology of Native American work out, *The Way We Lived*. **Stephanie Mills** lives in San Francisco and helps edit *CoEvolution Quarterly*. **Piro Caro** is an ex planner for the State of Illinois who has resided in the Sausalito houseboat community for 30 years. **Keith Abbott** lives in Berkeley and works as a tree cutter. His story is excerpted from *Mordecai of Monterey*, a novel in search of a good publisher. **Peter Coyote** is busy working on several films and is former head of the California Arts Commission. **Joseph Carey** was a photo-journalist in Vietnam and is now a restauranteur. **Herb Caen** has nothing to do with this book. **Stephen Kessler** edits Alcatraz Editions and lives in Santa Cruz. **Phil Frank** does *Travels with Farley*, a syndicated comic strip. **Bob Beatty** manages the Berkeley dump. **Bruce Boston** is the author of *Jackbird* (Berkeley Poets Coop Press) and is a retired furniture mover. He teaches the short story at JFK University in Orinda. **Lucia Berlin** lives in Berkeley. Her *Angels Laundromat* was recently published by Turtle Island Press. **Joan Schirle** plays Scar Tissue in *Intrigue at Ahpah*, a play she wrote for the Dell 'Arte Players. *Performance Anxiety* is their current exciting production. **Mort McDonald** works at Bug Press in Arcata. **Frank Kiernan** is a bartender in Oakland. **Ernest Callenbach** edits *Film Quarterly* for U.C. Press and is about to publish his new novel, *Ectopia Emerging*. **David Prowler** works for Trust For Public Land in San Francisco. **Peter Berg** is founder of Planet/Drum Foundation and edited *Reinhabiting A Separate Country*. **Jack Forbes** is a Native American who teaches at U.C. Davis. **Elizabeth Davis** is on the staff of the Holistic Childbirth Institute in San Francisco. **Michael Helm** has been the driving force behind *City Miner* (magazine and books) since 1976. He is a retired poet and sometimes journalist who recently appreciated C.M. Cioran's remark that "Ennui is the echo within us of time tearing itself apart." Or in his own words, "If not this, what else?"